SpringerBriefs in Applied Sciences and Technology

For further volumes:
http://www.springer.com/series/8884

Dan Wei

Micromagnetics and Recording Materials

 Springer

Dan Wei
Department of Materials Science and Engineering
Tsinghua University Beijing
Shuangqing 100084
Beijing
China

ISSN 2191-530X   ISSN 2191-5318   (electronic)
ISBN 978-3-642-28576-9   ISBN 978-3-642-28577-6   (eBook)
DOI 10.1007/978-3-642-28577-6
Springer Heidelberg New York Dordrecht London

Library of Congress Control Number: 2012938234

Printed on acid-free paper

Springer is part of Springer Science+Business Media (www.springer.com)

Preface

The phrase "micromagnetics" was brought up by William Fuller Brown Jr. in 1958, now it has become mainstream theory for computational applied magnetism. Brown wrote the first book named as *Micromagnetics* in 1963, he summarized the "magnetization curve theory" and "domain theory" before the 1960s, which are still the main parts of today's micromagnetics. In micromagnetics, the magnetic materials are discretized into micromagnetic cells, the equation of motion for magnetic moments in cells is the Landau-Lifshitz equation, and the magnetostatic interactions among all cells are calculated following the spirit of Maxwell's equations.

In the 1980s, two main computational micromagnetic methods were developed: finite difference method (FDM) and finite element method (FEM). In this book, two improvements of the FDM-FFT method are introduced. In the calculation of M-H loops, microstructures of polycrystalline thin films are included, thus the magnetic properties of hard magnetic media, FeCo soft magnetic thin film, and TMR multilayers can be solved in the same frame. In the domain calculation of devices, analytical demagnetizing matrices of polyhedron cells are found, thus the improved FDM-FFT method can be used to solve domains in arbitrary-shaped devices.

The author would like to thank Prof. Zheng Yang and Prof. Fulin Wei in Lanzhou University, Dr. Kai-Zhong Gao in Seagate Technology, Prof. Jing Zhu and Prof. Bai-Xin Liu in my department. I would also like to thank my students Dr. Sumei Wang; Hong-Jia Li; Kai-Ming Zhang; Yi Wang and more; without their inspirations and contributions, this book is impossible.

Beijing, China, July, 2011   Dan Wei

Contents

Chapter 1
Exordium

Abstract This chapter will review the source of knowledge in basic science and industry for this book. The electromagnetism developed in eighteenth and nineteenth centuries are one of the forerunners for micromagnetics. The mathematical expression, such as vector algebra, of Maxwell's equations will be discussed in Chap. 2. The fundamental magnetism is usually treated as part of the Solid State Physics. The quantum physics of materials and parameters used in micromagnetics will be given by the fundamental magnetism, thus this part will also be reviewed here. From the very beginning of the micromagnetics, it has been focusing on the magnetic recording industry. Therefore the key developments of the magnetic materials and devices utilized in magnetic information storage industry will be also summarized here.

Keywords Electromagnetism · System of units · Quantum spin · Magnetic recording media · Magnetic heads

Magnetic materials were discovered since the early history before the century. The "magnetite" ore was named after the province Magnesia in ancient Greek. The same material was utilized to build the primitive compass in the "point-south chariot" in ancient China. However, the understanding of magnetic materials takes a very long time. The structure and the ferrimagnetic nature of the magnetite were clarified only after Louis Neél's work in the 1950s.

In Maxwell's book *A Treatise on Electricity and Magnetism* published in 1873 [1], he discussed the "induced magnetization" phenomena from Chaps. 4 to 6 in Part III Magnetism, discussing how the magnetization in a magnet was varied by the external field. The earliest theory of "induced magnetization" was given by Walther Eduard Weber in 1852; in his theory, a linear magnetization near zero field and a saturation at large field were derived; however no hysteresis was included and discussed. Today, this part of magnetism is called the Applied Magnetism or the Technical Magnetism, where the domains, hysteresis loops and permeabilities will be studied for magnetic materials.

D. Wei, *Micromagnetics and Recording Materials*, SpringerBriefs in Applied Sciences and Technology, DOI: 10.1007/978-3-642-28577-6_1, © The Author(s) 2012

Maxwell has pointed out in his book that the so called "soft magnetic iron" is soft mechanically, and the "hard magnetic iron" is also hard in strength. This is an important fact. The polycrystalline microstructure of a magnetic material is the key to understand these properties. It is well known that more defects in solids will cause higher mechanical strength; therefore the soft or hard magnetic properties of materials are determined not only by the magnetic nature of the crystalline phase in grains, but also by the characteristics of defects or amorphous phase at the grain boundary. In this book, the theory of Applied Magnetism, called Micromagnetics, will be introduced to understand the technical properties of magnetic materials, which should be built on the microstructure of solids.

In the nineteenth century, the electric power, the wired and wireless communication became the first batch of industries developed after the breakthrough of electricity and magnetism. The magnetic materials were served as the soft magnetic cores or the hard magnets in these industries. In the twentieth century, the information industry is the most important newly developed industry, which changes the life of all people [2]. In the information industry, two intrinsic properties of electrons, charge and spin, are the key to realize the processing and storage of analogy or digital signals. The magnetic materials used in the information storage industry are called "magnetic recording materials". This class of magnetic materials will be the main focus of this book. Physics is an experimental science. Micromagnetics not only has scientific value in applied magnetism, but also serves as part of the basic knowledge in research and development of magnetic recording industry since 1950s [3]. Today, more than 90% of electronic information is stored in computer hard disk drives. That is why this book is named as "Micromagnetics and Recording Materials".

1.1 Electromagnetism and System of Units

William Gilbert's book *De Magnete* at the end of sixteenth century was usually treated as the beginning of modern electromagnetism. He introduced the word "electricity" from the Greek word "amber"; and he inferred that the earth is a large magnet from the compass experiment. In 1644, Rene Descartes accepted Gilbert's idea and draw "lines" for earth magnetic field, as seen in Fig. 1.1; he also said that iron was formed inside the earth, thus the field lines tend to penetrate the iron.

The breakthrough of electromagnetism appeared within the 50 years from 1780s to 1830s. In 1832, the concept of the system of units was brought up by the great mathematician Carl Friedrich Gauss, who was studying the magnetic field of earth together with Wilhelm Eduard Weber in Göttingen. Gauss concluded that all force in experiments can be measured by motions, and only three units for motion description are necessary for this "absolute system of units": length, time and mass. All other units can be derived from these three basic units, as seen in Table 1.1. This was a neat and brilliant thought from a top mathematician, which provided a systematic solution for the quantitative calculations in physics.

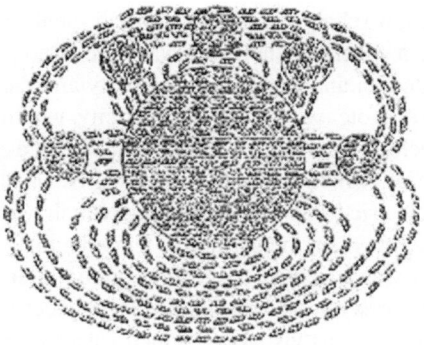

Fig. 1.1 Rene Descartes's drawing of earth magnetic field in 1644, the small balls represent magnetic materials like iron

Table 1.1 Basic quantities in electromagnetism, in the square brackets the respective unit is expressed by the base units in MKS or cgs system of units

Quantity	MKS units	cgs units	Unit conversion
Length l	m (meter)	cm (centimeter)	$1\,m = 10^2\,cm$, $1\,in = 2.54\,cm$
Mass m	kg (kilogram)	g (gram)	$1\,kg = 10^3\,g$, $1\,pd = 453.6\,g$
Time t	s (second)	s (second)	$1\,Hertz = s^{-1}$
Current I	A (ampere)	esa $[g^{1/2}\,cm^{3/2}/s^2]$	$1\,A = 3 \times 10^9\,esa$
Energy U, $\mathscr{E}$	J $[kg\,m^2/s^2]$	erg $[g\,cm^2/s^2]$	$1\,J = 10^7\,erg$
Force **F**	N $[kg\,m/s^2]$	dyne $[g\,cm/s^2]$	$1\,N = 10^5\,dyne$
Charge q	C [A s]	esu $[g^{1/2}cm^{3/2}/s]$	$1\,C = 3 \times 10^9\,esu$
Potential V, ψ	V $[kg\,m^2/s^3/A]$	esv $[g^{1/2}cm^{1/2}/s]$	$1\,V = \frac{1}{3} \times 10^{-2}\,esv$
Electric field **E**	V/m $[kg\,m/s^3/A]$	esv/cm $[g/cm^{1/2}/s]$	$1\,V/m = \frac{1}{3} \times 10^{-4}\,esv/cm$
Electric displacement **D**	C/m^2 $[A\,s/m^2]$	esu/cm^2 $[g/cm^{1/2}/s]$	$1\,C/m^2 = 3 \times 10^5\,esu/cm^2$
Polarization **P**	C/m^2 $[A\,s/m^2]$	esu/cm^2 $[g/cm^{1/2}/s]$	$1\,C/m^2 = 3 \times 10^5\,esu/cm^2$
Magnetic flux Φ	Wb $[kg\,m^2/s^2/A]$	Mx $[g^{1/2}cm^{3/2}/s]$	$1\,Wb = 10^8\,Mx$
Magnetic induction **B**	T $[kg/s^2/A]$	G $[g^{1/2}/cm^{1/2}/s]$	$1\,T = 10^4\,G$
Magnetic field **H**	A/m [A/m]	Oe $[g^{1/2}/cm^{1/2}/s]$	$1\,A/m = 4\pi \times 10^{-3}\,Oe$
Magnetization **M**	A/m [A/m]	emu/cm^3 $[g^{1/2}/cm^{1/2}/s]$	$1\,A/m = 10^{-3}\,emu/cm^3$

Gauss utilized mm, s, mg for the units of length, time and mass respectively. In 1881, the First International Conference of Electricians in Paris selected cm, g and s for the units of length, mass and time; after that, the Gauss units are also called cgs units. In 1893, the International Electrical Congress in Chicago adapted the SI units (Le Système International d'Unités), which has seven base units, and four of the base units related to electromagnetism are listed in the first part of Table 1.1. The SI units are also called MKS units, where m, kg and s are chosen for the units of length, mass and time. In the information industry, some units in the British system are still used now, which are also given in Table 1.1 [4].

The electricity became a science after the establishment of the Coulomb's law. In 1767, Joseph Priestley, a priest in Cheshire of England, found that inside an electrified conductor sphere there is no charge. Isaac Newton had spent years to build up calculus and proved that inside a spherical shell of mass there is no gravitational attraction. Thus Priestley suggested an analogy between gravity and electricity. From 1771 to 1781, Henry Cavendish wrote two papers of electricity, where he brought up several important concepts, including: (1) the dielectric constant of a capacitor or a material; (2) the electric potential, which he called the "degree of electrification"; (3) the inverse square law of electric force versus distance, which was related to the fact that the static charge always stays on the surface of a conductor.

The Coulomb's law was accepted by general public after Charles Augustine de Coulomb performed a series of experiments involving electric charges in 1785. The torsion balance experiment was visual and much easier to be understood than the screening effect of conductors. The $1/r^2$ force was also found independently by Coulomb in 1786. The system of units of electricity were built up gradually. Gauss utilized the Coulomb's law to define the measure of charge in cgs units:

$$F_{12}^{cgs} = \frac{q_1 q_2}{r^2} = 10^5 F_{12}^{MKS}; \quad F_{12}^{MKS} = k_e \frac{Q_1 Q_2}{R^2} = \frac{Q_1 Q_2}{4\pi\varepsilon_0 R^2}; \quad (1.1)$$

The k_e is the "Coulomb constant", which is directly related to the vacuum permittivity $\varepsilon_0 = (36\pi)^{-1} \times 10^{-9}$ F/m. Therefore the unit conversion between MKS and cgs units $1\,C = 3 \times 10^9$ esu can be found by Eq. (1.1). In 1811, 1824 and 1839, Siméon Poisson, George Green and Carl Gauss respectively applied the potential theory developed for gravitation on the Coulomb's law, and later for the Gauss' law for magnetism, which will be discussed further in next chapter.

The electric field $\mathbf{E}$ and the magnetic field $\mathbf{H}$ are directly related to the charge density ρ and the current density $\mathbf{j}$ respectively. The relationship between $\mathbf{E}$ and ρ is just the Coulomb's law. The hidden connection between the current and the magnetic field was explored by Hans Christian Oersted in 1820 when he prepared a course experiment. The mathematical expression of Oersted's experiment was given by the theorem of Ampere's closed loop in 1823.

The Coulomb's law provided an electrostatic unit of charge quantity. In 1852, Weber measured the force between two long wires at a distance of $r_{12} = 1$ cm; when the force between two wires per unit length was the same as the F_{12}^{MKS} with $R = 1$ cm in Eq. (1.1), the current in the wire provided another electromagnetic unit of charge quantity. This could be explained by the Ampere's law:

$$d\mathbf{F}_{12}^{MKS} = k_m I_2 d\mathbf{l}_2 \times \frac{I_1 d\mathbf{l}_1 \times \hat{r}_{12}}{r_{12}^2} \quad (1.2)$$

where the constant $k_m = \mu_0/4\pi = 10^{-7}$ H/m. Weber astonishingly found that the ratio between the electromagnetic unit over the electrostatic unit of charge quantity is just the speed of light $\sqrt{k_e/k_m} = c$, which provides a deep relationship between the MKS units of the permittivity and permeability:

Table 1.2 Electromagnetic quantities of materials, in the square brackets the respective unit is expressed by the base units in MKS or cgs system of units. The symbols $\bar{\varepsilon}_0$ and $\bar{\mu}_0$ represent the magnitude of ε_0 and μ_0 respectively

Quantity	MKS units	cgs units	Unit conversion
Capacitance C	Faraday $[A^2s^4/kg/m^2]$	cm [cm]	$1\,F = 9 \times 10^{11}$ cm
Permittivity ε	F/m $[A^2s^4/kg/m^3]$	1	$1\,F/m = 1/\bar{\varepsilon}_0 = 36\pi \times 10^9$
Inductance L	Henry $[kg\ m^2/A^2/s^2]$	s^2/cm $[s^2/cm]$	$1\,H = \frac{1}{9} \times 10^{-11}$ s^2/cm
Permeability μ	H/m $[kg\ m/A^2/s^2]$	1	$1\,H/m = 1/\bar{\mu}_0 = \frac{1}{4\pi} \times 10^7$
Resistance R	Ohm $[kg\ m^2/A^2/s^3]$	s/cm [s/cm]	$1\,\Omega = \frac{1}{9} \times 10^{-11}$ s/cm
Resistivity ρ	$\Omega{\cdot}m$ $[kg\ m^3/A^2/s^3]$	s [s]	$1\,\Omega \cdot m = \frac{1}{9} \times 10^{-9}$ s
Conductance G	Siemens $[A^2s^3/kg/m^2]$	cm/s [cm/s]	$1\,Sm = 9 \times 10^{11}$ cm/s
Conductivity σ	Sm/m $[A^2s^3/kg/m^3]$	1/s [1/s]	$1\,Sm/m = 9 \times 10^9$ 1/s

$$\varepsilon_0\mu_0 = \frac{1}{c^2} \simeq \frac{1}{9} \times 10^{-16}\ s^2/m^2 \tag{1.3}$$

Weber's measurement was so important that Maxwell repeated it in 1868–1869, and it became an accurate measurement method of speed of light. In Tables 1.1 and 1.2, it can be seen that the unit conversion often include a number "3", this is obviously related to the speed of light.

Faraday's law of electromagnetic induction, which is another connection between the electric and magnetic phenomena, was explored by Michael Faraday in 1831. Faraday did this experiment with an equipment: the Faraday's ring, made by soft magnetic iron with two winding coils. He found that the electric field and the magnetic field tend to encircle with each other. If the current in coil A wrapped on the Faraday's ring's was turned on, the magnetic field **H** in the ring suddenly varied, following the Ampere's law. The related variation of the magnetic flux density **B** in the ring was enlarged by large permeability of the soft iron material, finally another current was induced in coil B which was also wrapped on the ring.

In 1851, George Gabriel Stokes provided the mathematical form for the Faraday's law of electromagnetic induction, actually it also gave the differential form for the Ampere's theorem of closed loop. In 1862, following Faraday's idea, Maxwell introduced an extra term "displacement current" $\partial\mathbf{D}/\partial t$ to the current density **j** in the Ampere's theorem of closed loop, and this completed the discovery of the laws in electromagnetism since 1767.

In 1837, Faraday clarified the significant concepts of the "dielectric material" and the "(ferro)magnetic material". Faraday found the significant influence of dielectric materials on the static electric process, he thought that the electric field **E** provided an "electric stress" on dielectrics, and the electric displacement vector **D** thus varied with **E**. Faraday measured all existing elements at his time, and he surprisingly found that only Fe, Co, Ni are ferromagnetic. The MKS and cgs units of electromagnetic materials are listed in Table 1.2. If the materials are "linear", the relationship among fields, materials and source are as follows:

$$\mathbf{D} = \mathbf{E} + 4\pi\mathbf{P} = \varepsilon\mathbf{E} \quad \text{(cgs)} \qquad \mathbf{D} = \varepsilon_0\mathbf{E} + \mathbf{P} = \varepsilon_0\varepsilon_r\mathbf{E} \quad \text{(MKS)} \quad (1.4)$$

$$\mathbf{B} = \mathbf{H} + 4\pi\mathbf{M} = \mu\mathbf{H} \quad \text{(cgs)} \qquad \mathbf{B} = \mu_0(\mathbf{H} + \mathbf{M}) = \mu_0\mu_r\mathbf{H} \quad \text{(MKS)} \quad (1.5)$$

$$\mathbf{j} = \sigma\mathbf{E} \qquad\qquad\qquad \text{(cgs and MKS).} \qquad\qquad\qquad\qquad (1.6)$$

The differential form of the Ohm's law, which was originally derived by Georg Simon Ohm in 1826, is given in Eq. (1.6).

One of the main targets of this book is to clarify the nonlinear and multi-value relationship among the magnetization $\mathbf{M}$ and the external field $\mathbf{H}$ in magnetic materials. In ferromagnetic materials, the permeability μ or μ_r is not well defined, and the magnetic property has to be stated by the hysteresis or M–H loop.

1.2 Spin and Fundamental Magnetism

Micromagnetic theory is beyond the range of electromagnetism. The characteristics of ferromagnetic crystals, such as structure, saturation magnetization, anisotropy and exchange, are important and actually the inputs in micromagnetics. These characteristics were understood only after the breakthrough of Quantum Mechanics and Solid State Physics in the early twentieth century.

In 1840s, Faraday measured magnetic properties of a large amount of matter and he found that, among the known elements at that time, only Fe, Co, Ni crystals are ferromagnetic. The remanent magnetization of Fe, Co or Ni is mainly contributed by the spins of 3-d electrons in these crystals. As Laue said, a great discovery was always a nature merge of different branches of knowledge that seemed to be irrelevant to one another before [5]. The electron spin was discovered by analyzing the spectroscopy of gases with quantum principles.

In 1896, Pieter Zeeman discovered the magnetic splitting of spectral lines. In 1916, Arnold Sommerfeld modified circular orbits in the Bohr model to elliptical orbits to explain the Zeeman splitting. In an external magnetic field, the energy level would split due to the Zeeman energy:

$$E = -\mu_a \cdot \mathbf{H} = \mu_B H l_z ; \quad (l_z = l, l-1, ..., -l) \qquad (1.7)$$

where the Bohr magneton $\mu_B = e\hbar/2mc = 9.27 \times 10^{-21}$ erg/G is the quantum of moment, which was obtained based on the Ampere's assumption of molecular current magnetic moment in 1821 and a quantized angular momentum $l_z\hbar$ along the external magnetic field after Sommerfeld. The normal Zeeman effect must have odd number of spectral lines, as seen in Eq. (1.7); however, in a sodium gas, the anomalous Zeeman effect was observed with even number of splitting lines, as shown in Fig. 1.2a, which could not be explained then.

Further studies on spectra of atoms revealed that the single lines what were originally thought to be were actually closely spaced pairs of lines, implying extra degree of freedom for electrons. In early 1925, before the birth of Quantum Mechanics,

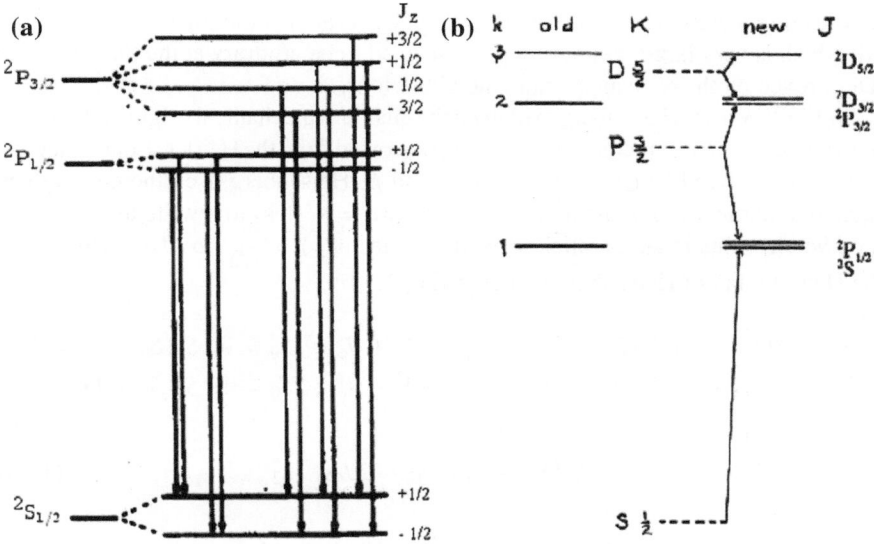

Fig. 1.2 Spectra of atoms for discovery of electron spin. **a** Anomalous Zeeman splitting of sodium atoms; **b** the old ($l = k = 0, 1, 2$) and the new scheme of energy levels in $n = 3$ orbit of hydrogen by Goudsmit and Uhlenbeck

the Pauli's repulsive principle was published. The Dutch physicists George Uhlenbeck and Samuel Goudsmit realized that the Pauli principle became easier to understand when introducing different quantum numbers. Thus a fourth quantum number $\sigma = s/\hbar = \pm\frac{1}{2}$ for spin is introduced for an electron together with the quantum numbers n, l, l_z of coordinates (r, θ, ϕ); and the doublets in the hydrogen spectra can be explained, as seen in Fig. 1.2b. The discovery of the spin is a fundamental breakthrough, which explains many phenomena, such as the periodic table, and the basic magnetic properties discussed here.

The fundamental theory of magnetism, especially ferromagnetism, is not completed up to date, even if great progresses had been made in the twentieth century [6]. In the early twentieth century, as an approximation, a solid was treated as a set of ions and valence electrons, where the interactions among different parts were neglected. In metals, the localized electrons in ions contribute much more to ferromagnetic properties than the valence electrons. This method is straightforward and has clear picture of physics, thus even now it is still often used.

In 1907, Pierre-Ernest Weiss, who was Louis Néel's mentor, suggested the concept of an "intrinsic magnetic field" H_E to explain ferromagnetism:

$$H_E = \lambda_E M = \lambda_E n \mu_a; \quad \mu_a \cdot \mathbf{H} = g S \mu_B H_E(0) \simeq k_B T_c \qquad (1.8)$$

where n is atomic density, atomic magnetic moment $\mu_a = g S \mu_B \simeq 2 S \mu_B$, and total spin S is used instead of $J = |L \pm S|$ due to the quenching of the orbital momentum in transition metals. The Weiss field $H_E(0)$ at $0\,\mathrm{K}$ is at the order of 10^7 Oe in Fe, Co, Ni, which is very large. This assumption seemed to be arbitrary at the time, but was later proved by the exchange among neighbor ions.

In 1928, Werner Heisenberg explained the quantum mechanical origin of the Weiss field by the exchange interaction, a concept borrowed from the Heitler–London theory of a covalent bond between two atoms α and β. Heisenberg used the eigenvalues (quantum number S_0) of the total spin operator $\mathbf{S} = \mathbf{s}_1 + \mathbf{s}_2$ to rewrite the energy of the two electrons in antiparallel or parallel states $|\Psi_\pm\rangle = \frac{1}{\sqrt{2}}\left(|\alpha_1\beta_2\rangle \pm |\beta_1\alpha_2\rangle\right)$ as the Hamiltonian of Heisenberg model in Eq. (1.10):

$$\varepsilon_+ = \langle\Psi_+|\mathscr{H}_0 + V|\Psi_+\rangle = (\varepsilon_\alpha + \varepsilon_\beta) + \bar{V} + J_e \; ; \; S_0 = 0, \; S_0(S_0 + 1) = 0$$
$$\varepsilon_- = \langle\Psi_-|\mathscr{H}_0 + V|\Psi_-\rangle = (\varepsilon_\alpha + \varepsilon_\beta) + \bar{V} - J_e \; ; \; S_0 = 1, \; S_0(S_0 + 1) = 2$$
$$(1.9)$$

$$\varepsilon = (\varepsilon_\alpha + \varepsilon_\beta) + \bar{V} - J_e\left(\mathbf{S}^2 - 1\right) = \varepsilon_0 - 2J_e\mathbf{s}_1 \cdot \mathbf{s}_2 \qquad (1.10)$$

In ferromagnetic materials, the "extra potential" $V = \frac{Z^2 e^2}{r_{\alpha\beta}} + \frac{e^2}{r_{12}} - \frac{Ze^2}{r_{1\beta}} - \frac{Ze^2}{r_{2\alpha}}$ other than the Hamiltonian of isolated atoms can be positive, then the exchange coupling $J_e = \langle\alpha_1\beta_2|V|\beta_1\alpha_2\rangle$ is positive and the ground states of the nearest neighbor atomic spins are parallel. If each atom has z neighbors, the total exchange energy by Eq. (1.10) must be equivalent to the Zeeman energy of the Weiss field at $0\,\mathrm{K}$:

$$E_{ex} = -2zJ_eS^2 \simeq -g\mu_B H_e(0)S \qquad (1.11)$$

where the Weiss field $H_e(0)$ at $0\,\mathrm{K}$ is a mean field and it is proportional to the magnetization $H_e(0) = \lambda_E n\mu_a \simeq \lambda_E ng\mu_B S$. Then the Heisenberg exchange energy constant J_e can be directly related to the Curie temperature T_c as:

$$J_e = \lambda_E \frac{n(g\mu_B)^2}{2z} = \frac{3k_B T_c}{n(g\mu_B)^2 S(S+1)} \frac{n(g\mu_B)^2}{2z} = \frac{3k_B T_c}{2zS(S+1)} \qquad (1.12)$$

where the Weiss parameter is proportional to the Weiss temperature as $\lambda_E n\mu_a^2 = 3k_B\theta \simeq 3k_B T_c$. The exchange energy J_e at $0\,\mathrm{K}$ is at the order of $0.1k_B T_c$, $0.01\,\mathrm{eV}$ or 10^{-14} erg in Fe, Co, Ni close-packed ferromagnetic metals.

After 1960s, the computational method of energy bands based on the Density Functional Theory (DFT), brought up by Kohn, Hohenberg and Sham, gradually became the mainstream of the fundamental theory of magnetism. The electronic structure of ferromagnet has splitting up- and down-spin energy bands. Some basic quantities, such as the saturation magnetization at $0\,\mathrm{K}$ in Table 1.3, are calculated with a higher accuracy compared to $M_s \simeq ng\mu_B S$ in the early theories. The first principle theory of tunneling magnetoresistivity (TMR) is also successful [7]. However, the calculations of fundamental magnetic parameters can not be very accurate for practical magnetic alloys used in recording industry, due to the complexity in microstructures such as substitutional disorder and other defects.

Table 1.3 Characteristics of important ferromagnetic crystals

Quantity	Fe	Co	Ni	Gd	Dy	Ni_3Fe	FeCo
M_s/(emu/cc) at 0 K	1,752	1,446	512	2,060	2,920	–	–
M_s/(emu/cc) at 300 K	1,707	1,400	485	–	–	1,007	1,950
μ_a/μ_B at 0 K	2.2	1.71	0.606	7.63	10.2	0.68/3	2.5
T_c/K	1,043	1,388	627	293	85	890	1,256
Structure at 0 K	BCC	HCP	FCC	HEX	HEX	FCC	BCC
Easy axis	⟨100⟩	[001]	⟨111⟩	Tilted	Rotate	⟨111⟩	⟨100⟩

1.3 Magnetic Recording Materials and Devices

Ferromagnetic materials, usually just called magnetic materials, are classified as permanent magnets, soft magnetic materials, and magnetic recording materials, which provide static magnetic fields, magnetic field paths, and those used in magnetic recording industry, respectively. In this section, thin film magnetic materials and devices used in computer hard disk drives will be the main focus.

The magnetic recording technology started in 1898 when Danish engineer Valdemar Poulsen invented an audio recorder "telegrafon" using piano wire as the magnetic medium and the electromagnetic coil as the read/write head. In the past 100 years, there were four generic magnetic recording products having significant influence on people's life: (1) the Magnetophon audio recorder, developed by BASF and AEG, Germany, in 1933; (2) the quadruplex video recorder, developed by Ampex, USA, in 1956; (3) the Random Access Memory Accounting and Control (RAMAC) disk file, which is called computer hard disk drive today, developed by IBM, USA, in 1956; (4) the diskette (computer floppy disk), developed by IBM in 1967. The breakthrough of magnetic recording technology is given in Table 1.4. In 2009, the worldwide market of electronic data storage is shared by hard disk drive (HDD) (128 EB), flash (7.8 EB), solid state disk (0.7 EB) and other (<10 EB). Therefore HDD is by far the most important information storage technology. It is also natural that the information processing and storage technology root in the charge and spin intrinsic characteristics of electrons, respectively.

The technologies utilized in HDD cover a wide range of disciples in engineering, among them, three most important technologies are: (1) magnetic devices; (2) tribology and servo; (3) signal processing. The Winchester disk structure in Fig. 1.3 has been used for HDD since 1970s; the two key classes of magnetic devices are media and heads, designed for information storage and read/write respectively.

The key technologies of magnetic recording listed in Table 1.4 mostly belong to the category of magnetic devices, except the hydrodynamic air bearing support of head (tribology), the PD and PRML channels (signal processing). The products were designed for audio, video entertainments or data storage. The first product of data storage was the Drum Recorder made for military usage by a company Engineering Research Associates (ERA) in USA. Today, the audio, video and data storage applications have been integrated into the computer hard disk drives.

Table 1.4 Significant inventer, technology breakthrough and related products of magnetic recording system in the last 100 years

Inventer	Technology	Product
1931, Fritz Pfleumer	Magnetic tape on paper	
1933, Eduard Schüller, AEG	Ring head	Magnetophon
1947, Brush Development	γ-Fe_2O_3 data storage tape	Drum Recorder, ERA
1951, IBM	NRZI format, PD channel	Data storage tape drive
1953, IBM	Air-bearing support of head	
1955, Ray Dolby, Ampex	Rotary drum head	Video Recorder
1956, IBM	γ-Fe_2O_3 coating hard disk	RAMAC, density 2 Kb/in^2
1966, IBM	PRML channel	
1967, IBM	Isotropic γ-Fe_2O_3 particulate media	Diskette
1971, Robert Hunt, Ampex	Magnetoresistive (MR) reader	
1976, Shun-ichi Iwasaki	CoCr perpendicular media, SPT head	
1979, IBM	Thin film inductive (TFI) head	1979, Hard Disk 8 Mb/in^2
1981, Robert Potter, Lanx	CoCr thin film disk media	1991, Hard Disk 90 Mb/in^2
1994, IBM	GMR reader on longitudinal media	1996, Hard Disk 1 Gb/in^2
2004, Seagate	TMR reader on perpendicular media	2005, Hard Disk 200 Gb/in^2

NRZI non-return-to-zero-inverse, *PD* peak detection, *PRML* partial response maximum likelihood, *SPT* single pole type, *GMR* giant-magnetoresistive, *TMR* tunneling-MR

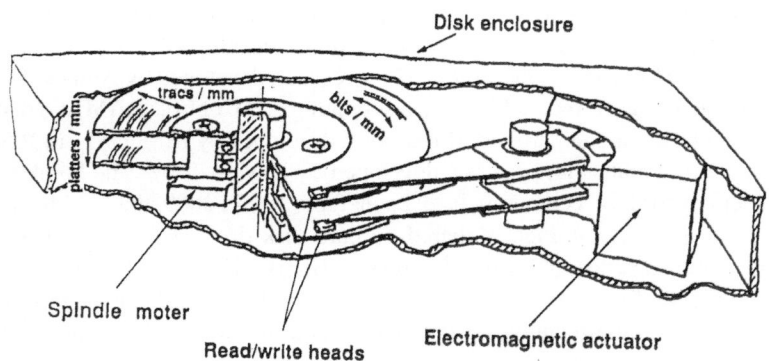

Fig. 1.3 Winchester disk structure invented by IBM in 1970s, which is continue to use up to now

The thin film technology in magnetic recording industry was borrowed from the semiconductor industry. Fairchild Semiconductor pioneered the planar process on silicon wafer developed by Jean Hoerni in 1958. The thin film head, introduced by IBM in 1979, represented a significant advance over the previously used magnetite ring head. The photolithographic and film deposition techniques was similar to the ones for fabricating solid state circuits in semiconductor industry. The thin film materials and devices in HDD will be discussed in following sections.

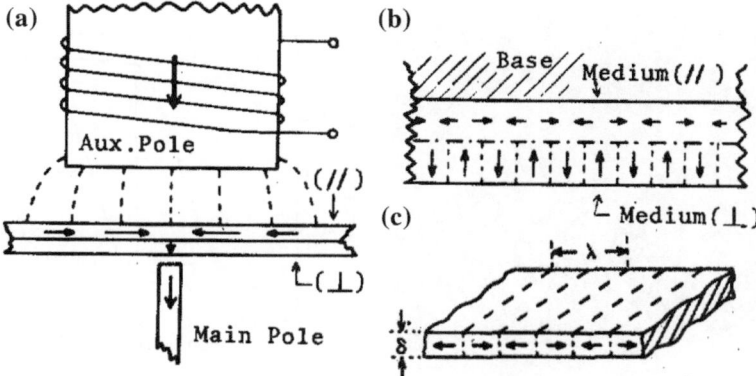

Fig. 1.4 Iwasaki's design of magnetic recording. **a** Effect of Fe–Ni soft-underlayer (SUL) recorded by a Single-Pole-Type (SPT) head with a main pole and an auxiliary pole; © [1979] IEEE. Reprinted, with permission, from Ref. [8]; **b** base, SUL and recording layer in a perpendicular thin film medium; © [1979] IEEE. Reprinted, with permission, from Ref. [8]; **c** longitudinal thin film medium; © [1980] IEEE. Reprinted, with permission, from Ref. [9]

1.3.1 Thin Film Media

Fritz Pfleumer's "sounding paper" invented in 1931 was the first generation of magnetic recording media. He glued pulverized iron particle onto a stripe of paper, creating the first magnetic tape. This type of particulate media were used for a long time, until the takeover of thin film disk media in 1990s. In the development of Magnetophon from 1931 to 1940, BASF improved the base film and medium respectively for better mechanical and magnetic properties: the substrate evolved from paper to cellulose acetate and plastic-based PVC; the magnetic particles such as carbonyl iron $Fe(CO)_5$, magnetite Fe_3O_4 and γ-Fe_2O_3 were used successively.

The data storage media built on aluminum or glass hard disks also had three generations. The γ-Fe_2O_3 particulate disk coating were used for the first 25 years, from 1956 to 1981, in all hard disk products. The longitudinal CoCrPt thin film disk media were used for the next 25 years, from 1981 to 2006. In the first 50 years of HDD, the areal density increased by 10^8 times, which was a miracle and symbolized HDD, together with CPU, as the most complicated high-tech products. A switch to the perpendicular CoCrPt-oxide thin film disk media happened in hard disk product in 2006, and these are still the disk media used today.

The thin film media was originally designed to realize the perpendicular recording, brought up by Professor Shun-ichi Iwasaki in Tohoku University, Japan, in 1975. The mode of magnetic recording had always been "longitudinal", since the beginning of magnetic recording in 1898, i.e. the averaged magnetic moment of particulate grains was in the longitudinal direction of the medium, as seen in Fig. 1.4c. Iwasaki thought that, the magneto-static field inside a longitudinal bit is in the opposite direction of the averaged magnetic moment; therefore, when the areal density is very

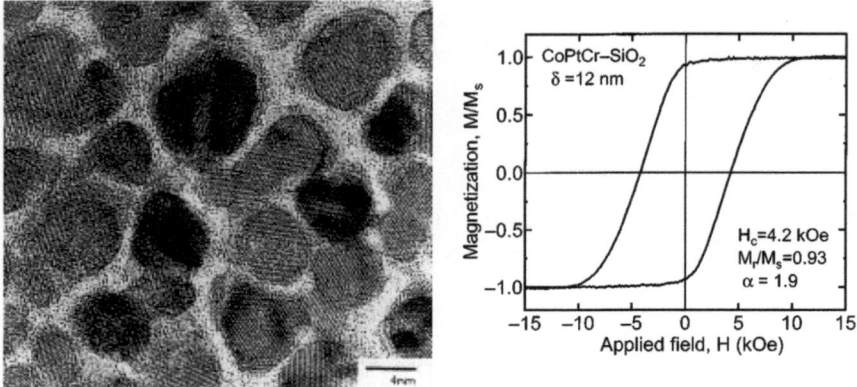

Fig. 1.5 Microstructure and M–H loop of CoCrPt-SiO$_2$ perpendicular recording media. © [2002] IEEE. Reprinted, with permission, from Ref. [10]

high, the longitudinal mode would finally fail due to the demagnetization effect. On the contrary, in a certain bit of perpendicular medium, the magneto-static field from a neighboring bit is in the same direction of the magnetic moment; thus the perpendicular mode is the mode for high density recording. In 1976, Iwasaki and his student Ouchi undesignedly discovered the CoCr perpendicular thin film media when they studied the magneto-optic media, and they added the Fe–Ni soft-underlayer (SUL) underneath the CoCr recording layer (RL), to enhance the perpendicular write field in RL, as seen in Fig. 1.4.

Iwasaki's idea was correct but in advance, since in 1975, the bit length B was much larger than the medium thickness δ in the longitudinal particulate disk media, and the demagnetization effect was still tolerable. Thirty years later, the bit length B was decreased to 25 nm, and became comparable to the thickness δ, which reached the transition point from longitudinal to perpendicular recording (Fig. 1.5).

Historically, it is the longitudinal but not perpendicular CoCr metallic thin film media were first applied in hard disk products. In May 1980, influenced by Iwasaki's work on CoCr media [8], Robert Potter resigned from IBM and started up a new company called Lanx. Potter and his partner Niel Heiman, who was also an engineer resigned from IBM, started to build up the magnetron sputtering vacuum deposition equipment for thin film disk media on aluminum substrate with a stellite (CoCrW) nonmagnetic underlayer, a CoCr magnetic layer and a carbon overcoat. By the end of 1980s, the 5.25 and 3.5 in longitudinal disk media on Al–Mg substrate, with a 5 μm NiP sublayer (controlling surface roughness), a 50 nm Cr underlayer (with (200) texture), a 30 nm CoCrPtTa magnetic layer (with (11$\bar{2}$0) texture) and a 10 nm carbon overcoat, became the universal design in hard disk industry. The key success of longitudinal CoCrX thin film media is the segregation of the nonmagnetic Cr-rich phase on the grain boundary naturally formed in the sputtering process, which ensures the relatively independent switching of magnetic grains and thus largely reduces the medium noise and increases the signal-to-noise ratio (SNR).

The CoCr perpendicular media could not be used in product before the year 2000 because the inter-grain exchange was too large. The CoCrX alloy still has the HCP structure, and its c-axis is perpendicular-to-plane by introducing an orientation control Ru underlayer. When the c-axis with sixfold rotational symmetry is perpendicular, the crystal lattices in neighbor magnetic grains are easy to match with one another due to the parallel c-axes; thus the density of defects at grain boundary would be much lower than that in longitudinal media, and the nonmagnetic Cr segregation can not be naturally formed. In 2002, Yoshihisa Nakamura's group in Tohoku University added SiO_2 to the CoPtCr magnetic layer to enhance the grain isolation. Compared to CoCrPtB perpendicular media without oxide grain segregation, the saturation magnetization M_s is increased from 380 to 560 emu/cm^3 and the anisotropy energy K_1 is largely improved from about 1×10^6 to 4×10^6 erg/cm^3. As a result, the demagnetization is suppressed and SNR is largely increased compared to CoCrPtB, which made this CoPtCr-SiO$_2$/Ru/SUL perpendicular media practical and became the universal design of media in the HDD product after 2005.

1.3.2 Write and Read Heads

In audio or video recorder, the signal is analog; in data storage systems, the signal is digital. The head is an electric–magnetic signal converter, by which the analog or digital electric signal is recorded on a magnetic medium. Before 1991, the same head was used in both the read and write process, where the ring head and the thin film inductive (TFI) head represented the two generations of heads in this era. In 1991, IBM used the magnetoresistive (MR) head as the read head in disk drive, to solve problem of the signal decrease at higher recording density; thus the annual growth of areal density in HDD reached 60%. In 1994, IBM invented the giant-magnetoresistive (GMR) read head, the annual growth of HDD areal density even approached 100% after 1997. Both MR and GMR readers were designed for longitudinal recording. In 2004, Seagate developed the tunneling-magnetoresistive (TMR) read head. The SPT writer and TMR reader were designed for perpendicular recording, which enabled the areal density growth up to 1 Tb/in^2.

In 1933, Eduard Schüller joined AEG in Germany, and brought up the idea of ring head, the basis of all future magnetic recording heads, as seen in Fig. 1.6a. Schüller used a stack of silicon steel plates to reduce the eddy current. The most important design was the gap at the tip of the head, with this design, the linear density or the bit length B can be directly controlled by the gap length g. The surface near the gap was rounded and highly polished, such that the head-medium-spacing (HMS) d was largely reduced, which was also a key to have a large read back signal. The ring head is one of the most fundamental inventions in magnetic recording.

The TFI head invented by IBM in 1979 represented a major advance over the heads of early disks. From the topological point of view, the TFI head was equivalent to the ring head, as seen in Fig. 1.6b; however, the TFI head was fabricated with a new level of precision and control of the gap length g, by using the photolithographic and

(b) IBM 3370 film head ⟶

  (A) magnetic layers
  (B) pole tips
  (C) conductor turns
  (D) gap layer
  (E) insulation layer

(a)

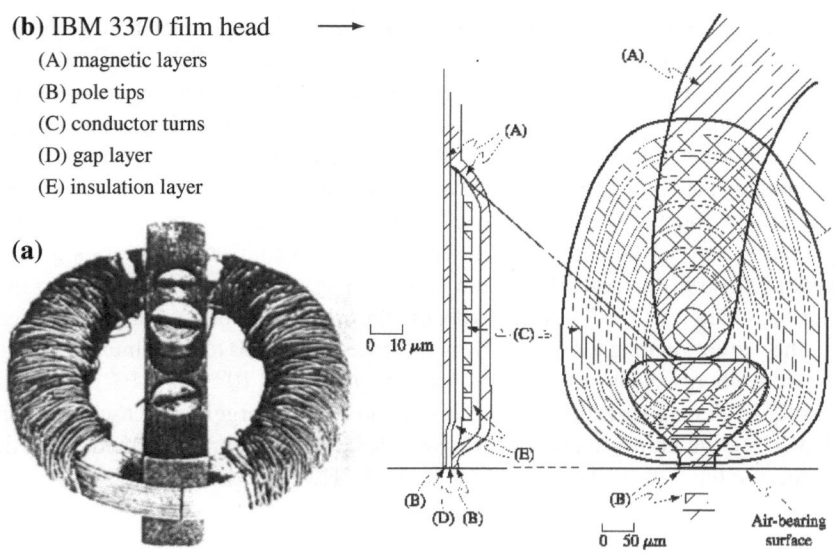

Fig. 1.6 Inductive head. **a** Ring head invented by Eduard Schüller in 1933 (Courtesy of AEG); **b** TFI head invented by IBM in 1979; © [1996] IBM. Reprinted, with permission, from Ref. [11]

film deposition techniques similar to the semiconductor industries. Actually the thin film head program in IBM was started in 1964. The vacuum deposition of $1\,\mu m$-thick permalloy (NiFe) and the $2\,\mu m$-width lithography had become dependable technologies in 1968, at this year the thin film device group in IBM developed a one-turn thin film head, using a TiC–Al_2O_3 glass–ceramic substrate. In the IBM 3370 film head developed in 1979, the magnetic layers were still permalloy; the material in gap was amorphous AlO thin film, which could prevent corrosion. The film thickness of left tip, gap and right tip is 1.6, 0.6 and $1.9\,\mu m$, respectively. The head width is $38\,\mu m$. After the segment of TiC–AlO substrate, a IBM 3370 film head slide had a size of $4.0 \times 3.2 \times 0.85\,mm^3$.

Although the separation of read and write heads in HDD products was not realized until 1991, the idea had been brought up by Hunt in Ampex twenty years earlier. Hunt utilized the ring head as the writer, and a MR transducer as the reader, as seen in Fig. 1.7. Actually the magnetoresistive (MR) effect or anisotropic magnetoresistive (AMR) effect was discovered by Lord Kelvin in 1856, he found that the resistivity $\rho = \rho_\parallel$ is maximized when the current $\mathbf{J}$ is in the same direction of magnetization $\mathbf{M}$, and the resistivity $\rho = \rho_\perp$ is minimized when $\mathbf{J}$ is at $90°$ to $\mathbf{M}$. In 1975, McGuire and Potter in IBM explained the AMR effect by the anisotropic conductivity matrix σ_{ij} in the microscopic Ohm's law $J_i = \sigma_{ij} E_j$ due to the anisotropic scattering of electrons in an external electromagnetic field [13]. In a ferromagnetic soft magnetic thin film like Ni-alloy, the magnetic induction $\mathbf{B} = \mathbf{H} + 4\pi\mathbf{M}$ is almost parallel to $\mathbf{M}$ in a small external field $\mathbf{H}$, therefore the resistivity is dependent on the angle between the current $\mathbf{J}$ and the magnetization $\mathbf{M}$.

Fig. 1.7 Composite head
of video tape drive invented
by Hunt of Ampex, with
a ring head writer and a
MR reader; © [1971] IEEE.
Reprinted, with permission,
from Ref. [12]

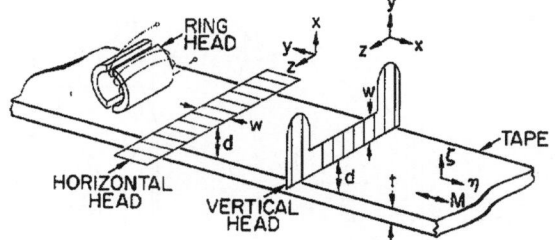

Actually, just by Hunt's vertical MR head design in Fig. 1.7, the signal had a degeneracy in a $+H_y$ or a $-H_y$ field from the tape. In 1985, IBM developed a MR or AMR head for the data storage tape system, where a soft adjacent layer (SAL) was added next to the vertical MR sensor, as seen in Fig. 1.8a, and two soft magnetic shields were put at both sides of MR-SAL layers to increase the resolution of bits along the track. When the sense current I is zero, the magnetization **M** in both the MR and SAL layers is in the cross-track z-direction. When a nonzero I is applied in the MR layer, the magnetic field of I rotates the **M** in SAL to $-y$ direction; the negative M_y in SAL creates a demagnetizing field and will bias the **M** in MR layer to $\theta_0 \simeq +45°$. This design ensured the linear readout near H_{ext} in MR head, and the MR voltage was solved with self-consistent micromagnetics by Smith [14]:

$$\Delta V_{MR} = JW\Delta\rho_{MR}\langle\cos^2\theta - \cos^2\theta_0\rangle \qquad \tan\theta \simeq \frac{H_{ext} + H_{bias}}{H_s} \qquad (1.13)$$

where J is the current density, W is the width of reader, and $\Delta\rho_{MR} = \rho_\parallel - \rho_\perp$ is the maximum variation of resistivity in the AMR effect. The angle θ between **M** and **J** should be solved self-consistently by considering the effect of the external field H_{ext}, the bias field H_{bias} from SAL and the shape anisotropy field H_s of the MR layer.

The design of MR head was already quite close to that of GMR head, as shown in Fig. 1.8; that might be part of the reason why IBM could transform the fundamental studies of GMR multilayers to GMR head products within 10 years. In 1960s, Albert Fert found that the resistivity of $Ni(Co_{1-x}Rh_x)$ alloys largely increase with the doping rate x. He explained this phenomena by a conducting channel with ABABAB... atoms. The total conductivity can be explained by Nevil Mott's two current model of spin dependent mobility brought up in 1936:

$$\rho^{-1} = \rho_\uparrow^{-1} + \rho_\downarrow^{-1} = (\rho_\uparrow^A + \rho_\uparrow^B)^{-1} + (\rho_\downarrow^A + \rho_\downarrow^B)^{-1} \ . \qquad (1.14)$$

If $\rho_\uparrow^A < \rho_\downarrow^A$ but $\rho_\uparrow^B > \rho_\downarrow^B$, the total resistivity ρ must increase with higher doping rate of B atoms, due to the increase of $\rho_\uparrow$ and $\rho_\downarrow$ in both spin channels, that is the case for $Ni(Co_{1-x}Rh_x)$ alloys. In 1986, Peter Grünberg, who studied Brillouin light scattering (BLS) of magnetic semiconductors EuO and EuS since 1970s in Jülich Germany, visited Argonne National Lab in the US and proved the antiferromagnetic

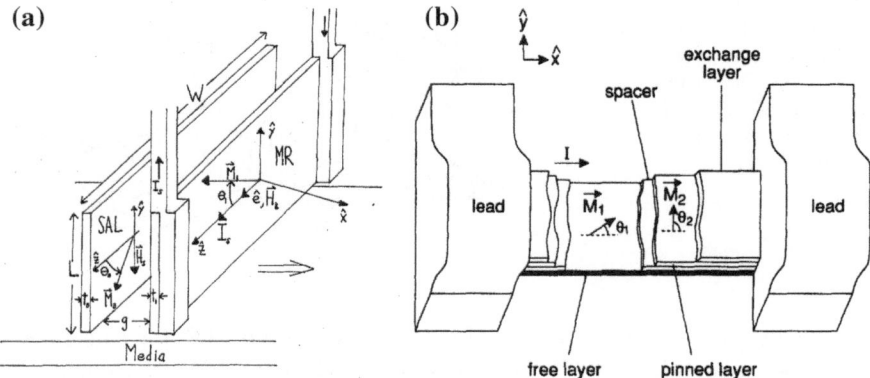

Fig. 1.8 Designs of MR and GMR read heads. **a** Mechanism of MR head with a MR and a SAL layer by Smith from Eastman Kodak; © [1987] IEEE. Reprinted, with permission, from Ref. [14]; **b** spin valve GMR multilayer sensor and leads in a GMR head invented by IBM; © [1994] IEEE. Reprinted, with permission, from Ref. [15]

coupling of two Fe layers by BLS in a high quality Fe/Cr/Fe samples provided by M. Brodsky. Albert Fert saw this result and realized that the two ferromagnetic layers in Fe/Cr/Fe can play the role of A and B atoms in the two current model. Using the MSE technique in Thomas-CAF, Albert's group prepared $(Fe/Cr)_n$ superlattice and discovered the GMR effect in 1988 [16]. Peter Grünberg discovered the GMR effect in Fe/Cr/Fe trilayers almost at the same time [17]. This GMR effect can still be interpreted by Eq. (1.14): in zero external field, due to the RKKY interaction [18], magnetization in layer A and B are anti-parallel with a Cr-layer thickness $t \sim 6\,\text{Å}$, then both spin channels $\rho_\uparrow$ and $\rho_\downarrow$ are high for spin up or down conducting electrons, which corresponds to a high resistivity state; in a large external field, layer A and B is magnetically parallel, either $\rho_\uparrow$ or $\rho_\downarrow$ is small for spin up or spin down conducting electrons, which reveals a low resistivity state.

On the road to develop GMR head, the spin-valve structure GMR multilayer was a key invention. In 1991, Parkin et al. in IBM prepared a NiFe(60 Å)/Cu(25 Å)/NiFe (30 Å)/FeMn(70 Å) spin valve [19]. The first NiFe layer is a free layer and the second NiFe layer is pinned by the antiferromagnetic FeMn layer. The conducting current mostly passes the Cu layer. In 1994, the GMR head was developed based on the spin valve sensor [15], where the free layer (FL) is similar to the MR layer and the pinned layer (PL) plays the role of the SAL layer in AMR head, as seen in Fig. 1.8. By controlling the texture of FeMn and the deposition conditions, the magnetization M_2 in PL has an angle $\theta_2 = 90°$ and M_1 in FL is horizontal with $\theta_1 = 0°$ in zero external field, which assures a linear readout:

$$\Delta V_{GMR} = J W \Delta \rho_0 \, \cos(\theta_1 - \theta_2) \simeq J W \Delta \rho_0 \sin \theta_1 \qquad (1.15)$$

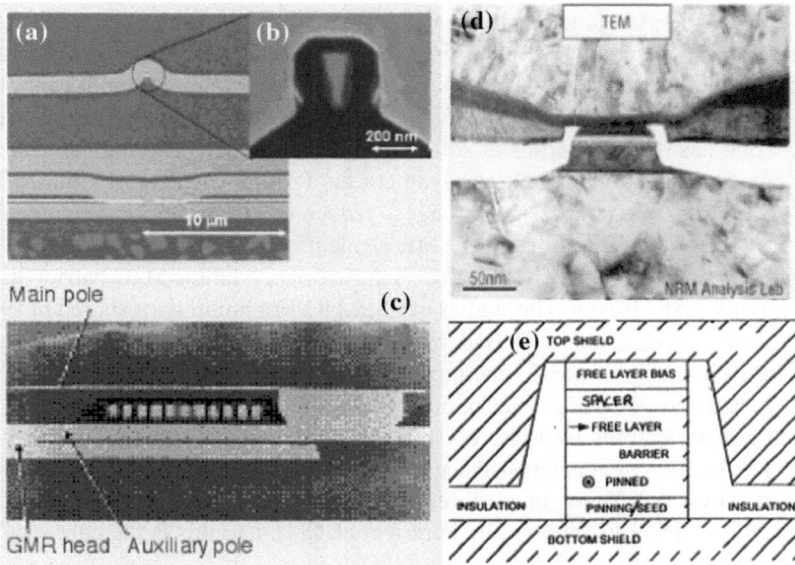

Fig. 1.9 Heads in perpendicular HDD: SPT writer from Hitachi; **a** air-bearing surface (ABS); **b** main pole structure; **c** down-track cross-section; © [2005] IEEE. Reprinted, with permission, from Ref. [20]. TMR reader from Seagate; **d** ABS view of 130G TMR head; **e** schematics of TMR multilayer structure; © [2004] IEEE. Reprinted, with permission, from Ref. [21]

where $\Delta\rho_{GMR} = 2\Delta\rho_0 \sim 5\%\rho$ in GMR head, much larger than $\Delta\rho_{MR} \sim 1.5\%\rho$ in AMR head. If the total resistance R of a GMR head is around 50 Ω and the sense current is 3 mA, the readout signal ΔV_{GMR} is around 1.5 mV in a 200 Oe external field, which corresponds to a 1% change of the total resistance.

The leaders in hard disk industry realized that the transform from longitudinal to perpendicular recording was necessary when the areal density of exceeded 100 Gb/in^2 in 2003. The choices of heads in perpendicular HDD were SPT writers and TMR readers, as shown in Fig. 1.9 respectively. The idea of SPT writer had been brought up by Iwasaki and Nakamura early in 1977. When SPT writer in Fig. 1.9a was finally industrialized, it was prepared by thin film technologies. The auxiliary pole and main pole tip have a spacing with one another, which form a loop for magnetic field together with the soft-underlayer (SUL) underneath the magnetic layer in the thin film medium. To improve the head field gradient, which is crucial for high density recording, a side shield or even a wrap-around shield is added around the main pole tip, as seen in Fig. 1.9a, b; furthermore, in present SPT heads, inclined surfaces are introduced on the main pole tip and the shields around the tip, to achieve even higher field gradient.

The tunneling magnetoresistance (TMR) of the magnetic tunneling junction (MTJ) had been discovered back in 1970s, but the effect was small and only appeared at low temperature. In 1995, large TMR effect at room temperature was discovered by two groups. Moodera discovered $\Delta\rho_{TMR} \sim 10\%\rho$ at 295 K in a CoFe/Al$_2$O$_3$/Co junction [22]; and Miyazaki observed $\Delta\rho_{TMR} \sim 20\%\rho$ in a Fe/Al$_2$O$_3$/Fe junction [23]. The first generation of TMR reader developed by Seagate in 2004, whose ABS view is shown in Fig. 1.9d, used a multilayer structure of Ta/NiFe/CoFe/AlO/CoFe/Ru/CoFe/PtMn/Ta. The combined free layer (FL) NiFe/CoFe has both properties of soft magnetic (from NiFe) and high atomic spin adjacent to barrier (from CoFe). The synthetic antiferromagnetic (SAF) pinned layer (PL) CoFe/Ru/CoFe is a design to largely reduce or eliminate the demagnetizing field from PL to FL in a finite-sized TMR element.

The next breakthrough of TMR effect is the use of crystalline MgO barrier instead of amorphous AlO barrier. The studies of FM/MgO/FM multilayers started in 1990s. In 2001 and 2004, Zhang and Butler in Oak Ridge National Lab in the US used first principle theory to predict the large MR $= \Delta\rho_{TMR}/\rho$ in trilayers Fe/MgO/Fe, Co/MgO/Co and FeCo/MgO/FeCo, due to the different selective scattering of spin up and down electrons by MgO crystals in parallel or antiparallel ferromagnetic layers [7]. In 2004, magnetoresistance of epitaxial Fe/MgO/Fe with a MR $= 247\%$ is reported by Shinji Yuasa; magnetoresistance of textured Co$_{84}$Fe$_{16}$/MgO/Co$_{70}$Fe$_{30}$ with a MR $\sim 150\%$ is reported by Parkin et al. [24], and this is the core structure used in TMR head after 2005.

In the later chapters in this book, micromagnetic theories will be brought up to analyze the magnetic properties of magnetic materials and devices introduced in this section. In Chap. 2, Maxwell's equations and Landau–Lifshitz–Gilbert equations will be introduced as a fundamental to micromagnetics. The Green's function method will also be discussed to calculate the demagnetizing matrix of an arbitrary polyhedron micromagnetic cell. In Chaps. 3 and 4, the calculations of M–H loops, static domain and dynamic domain motion would be analyzed respectively.

References

1. Maxwell, J.C.: A Treatise on Electricity and Magnetism (1873), translated by Ge G. into Chinese. Wuhan Press, Wuhan (1994)
2. Daniel, E.D., Mee, C.D., Clark, M.H.: Magnetic Recording—The First 100 Years. IEEE Press, New York (1999)
3. Bertram, H.N.: Theory of Magnetic Recording. Cambridge University Press, Cambridge (1994)
4. Wei, D.: Fundamentals of Electric, Magnetic, Optic Materials and Devices (in chinese), 2nd edn. Science Press, Beijing (2009)
5. von Laue M.: Geschichte der Physik (1950), translated by Fan D.N. and Dai N.Z. into Chinese. Commercial Press, Beijing (1978)
6. Wei, D.: Solid State Physics. Cengage Learning, Singapore (2008)
7. Zhang, X.G., Butler, W.H.: Large magnetoresistance in bcc Co/MgO/Co and FeCo/MgO/FeCo tunnel junctions. Phys. Rev. B **70**, 172407 (2004)
8. Iwasaki, S., Nakamura, Y., Ouchi, K.: Perpendicular magnetic recording with a composite anisotropy film. IEEE. Trans. Magn. **15**(6), 1456–1458 (1979)
9. Iwasaki, S.: Perpendicular magnetic recording. IEEE. Trans. Magn. **16**(1), 71–76 (1980)
10. Oikawa, T., Nakamura, M., Uwazumi, H., Shimatsu, T., Muraoka, H., Nakamura, Y.: Microstructure and magnetic properties of CoPtCr-SiO$_2$ perpendicular recording media. IEEE. Trans. Magn. **38**(5), 1976–1978 (2002)

11. Chiu, A., Croll, I., Heim, D.E., Jones, Jr, R.E., Kasiraj, P., Klaassen, K.B., Mee, C.D., Simmons, R.G.: Thin-film inductive heads. IBM J. Res. Dev. **40**(3), 283–300 (1996)
12. Hunt, R.P.: A magnetoresistive readout transducer. IEEE Trans. Magn. **7**(1), 150–154 (1971)
13. McGuire, T.R., Potter, R.I.: Anisotropic magnetoresistance in ferromagnetic 3d alloys. IEEE. Trans. Magn. **11**(4), 1018–1038 (1975)
14. Smith, N.: Micromagnetic analysis of a coupled thin-film self-biased magnetoresistive sensor. IEEE. Trans. Magn. **23**(1), 259–272 (1987)
15. Tsang, C., Fontana, R.E., Lin, T., Heim, D.E., Seriosu, V.S., Gurney, B.A., Williams, M.L.: Design, fabrication and testing of spin-valve read heads for high density recording. IEEE. Trans. Magn. **30**(6), 3801–3806 (1994)
16. Baibich, M.N., Broto, J.M., Fert, A., Nguyen van Dau, F., Petroff, F., Eitenne, P., Creuzet, G., Friederich, A., Chazelas, J.: Giant magnetoresistance of (001)Fe/(001)Cr magnetic superlattices. Phys. Rev. Lett. **61**(21), 2472–2475 (1988)
17. Binasch, G., Grünberg, P., Saurenbach, F., Zinn, W.: Enhanced magnetoresistance in layered magnetic structures with antiferromagnetic interlayer exchange. Phys. Rev. B **39**(7), 4828–4830 (1989)
18. Ruderman, M.A., Kittel, C.: Indirect exchange coupling of nuclear magnetic moments by conduction electrons. Phys. Rev. **96**, 99–102 (1954)
19. Dieny, B., Speriosu, V.S., Parkin, S.S.P., Gurney, B.A., Wilhoit, D.R., Mauri, D.: Giant magnetoresistance in soft ferromagnetic multilayers. Phys. Rev. B **43**(1), 1297–1300 (1991)
20. Okada, T., Nunokawa, I., Mochizuki, M., Hatatani, M., Kimura, H., Etoh, K., Fuyama, M., Nakamoto, K.: Newly developed wraparound-shielded head for perpendicular recording. IEEE. Trans. Magn. **41**(10), 2899–2901 (2005)
21. Mao, S., Linville, E., Nowak, J., Zhang, Z., Chen, S., Karr, B., Anderson, P., Ostrowski, M., Boonstra, T., Cho, H., Heinonen, O., Kief, M., Xue, S., Price, J., Shukh, A., Amin, N., Kolbo, P., Lu, P.L., Steiner, P., Feng, Y.C., Yeh, N.H., Swanson, B., Ryan, P.: IEEE. Trans. Magn. **40**(1), 307 (2004)
22. Moodera, J.S., Kinder, L.R., Wong, T.M., Meservey, R.: Large magnetoresistance at room termperature in ferromagnetic thin film tunnel junctions. Phys. Rev. Lett. **74**(16), 3273–3276 (1995)
23. Miyazaki, T., Tezuka, N.: Giant magnetic tunneling effect in Fe/Al$_2$O$_3$/Fe junction. J. Magn. Magn. Mater. **139**, L231–L234 (1995)
24. Parkin, S.S.P., Kaiser, C., Panchula, A., Rice, P.M., Hughes, B., Samant, M., Yang, S.H.: Giant tunnelling magnetoresistive at room temperature with MgO (100) tunnel barriers. Nat. Mater. **3**, 862–867 (2004)

Chapter 2
Maxwell Equations and Landau–Lifshitz Equations

Abstract This chapter will present the theoretical, mathematical and computational fundamentals for micromagnetics. The target of micromagnetics is to clarify the motion of magnetic moments in ferromagnetic materials and devices, which is described by the nonlinear Landau–Lifshitz equations, or the equivalent Landau–Lifshitz–Gilbert (LLG) equations. In the LLG equations, the time derivative of the magnet moment in a micromagnetic cell is controlled by the local effective magnetic field. The effective magnetic field contains the terms determined by the fundamental and applied magnetism in a magnetic material, including the external field, the crystalline anisotropy field, the exchange field, the demagnetizing field, and the magneto-elastic field. Among these field terms, the most time-consuming one in computation is the demagnetizing field, which will be calculated by the Green's function method following the Maxwell's equations.

Keywords Maxwell equations · Vector analysis · Demagnetizing matrix · Free energy · Landau–Lifshitz equations · History of micromagnetics

The development of micromagnetics was related to the computational science in an unusual way. In 1945, the great mathematician John von Neumann started a computer project in Princeton, to redesign the logic structure of the first computer ENIAC invented in 1944 by J. Presper Eckert and John Mauchly. Neumann brought up the idea of "program digital computer", including a processing unit, a data and program memory, a controller and an input/output device. The major problem in ENIAC was the lack of memory; thus Neumann suggested to build up a memory by vacuum tubes, and made the independent development of hardware and software possible. The storage capacity of vacuum tube memory was limited; therefore a magnetic recording tape drive "Uniservo" was built up as the external memory of the UNIVAC computer in 1948 by the first computer company Eckert–Mauchly Computer Corporation (EMCC). In 1945, Neumann already saw the great potential of computer to change the way of mathematical computation, especially for the nonlinear problems. At the heart of micromagnetics, there is such a nonlinear problem defined by the Landau–Lifshitz (LL) equations [1]; only after using the computational methods by Brown in 1960s [2], micromagnetics finally became dependable and served as the basic theory in magnetic recording industry.

The target of micromagnetics is to clarify the motion of magnetic moments in ferromagnetic materials and devices, which is described by the LL equations, or the equivalent Landau–Lifshitz–Gilbert (LLG) equations. In the LLG equations, the time derivative of the magnet moment $\hat{m}$ in a micromagnetic cell is controlled by the local effective magnetic field $\mathbf{H}_{\text{eff}}$, which varies along with $\hat{m}(\mathbf{r})$ and forms a high nonlinearity. The effective magnetic field $\mathbf{H}_{\text{eff}}$ contains the terms determined by the fundamental and applied magnetism, including the external field, the crystalline anisotropy field, the exchange field, the demagnetizing field, and the magneto-elastic field. Among these field terms, the most time-consuming one in computation is the demagnetizing field, which will be calculated by the Green's function method following the Maxwell's equations.

2.1 Maxwell Equations and Vector Analysis

In Maxwell's book *A Treatise on Electricity and Magnetism* [3], all equations of electromagnetic field were written in the component form. Similar to the mathematical development after Newtonian mechanics in the eighteenth century, there was also a period of mathematics development after Maxwell's equations. In late nineteenth century, Josiah Willard Gibbs and Oliver Heaviside developed the vector analysis, and Maxwell's equations were rewritten in the vector form, which is beautiful and even looks like an apocalypse, as commented by Laue [4].

Gibbs was awarded the first Ph.D. in Engineering in the US from Yale in 1863. He went to Europe in 1866, and spending a year each at Paris, Berlin, and Heidelberg. Maxwell's theory published in 1865 predicted the existence of electromagnetic waves moving at the speed of light. Hermann von Helmholtz was at Heidelberg at that time, and he was one of the first physicists in continental Europe who was interested in Maxwell's theory. Gibbs was influenced by Helmholtz's interests during his visit. In 1879, Helmholtz suggested his student Heinrich Hertz to test experimentally the Maxwell's theory of electromagnetism, which was finally done by Hertz later in 1886. In the same period, Gibbs invented vector analysis in Yale, and developed a theory of optics using his notation.

Oliver Heaviside, another independent inventor of vector analysis, abstracted Maxwell's set of equations to four equations in the vector form. Heaviside was a self-taught English electrical engineer, mathematician, and physicist. In 1873 Heaviside had encountered Maxwell's newly published book [3]. He felt it was great, greater and greatest with prodigious possibilities in its power, and determined to master the book. However he had no knowledge of mathematical analysis at that time. Finally in 1884, Heaviside reformulated Maxwell's 20 equations in 20 unknowns to the four vector equations using vector calculus. Those four vector equations, involving abstract mathematical terminology as curl and divergence, are formally called as "Maxwell's equations".

2.1.1 Vector Analysis

The first step to solve a problem is to describe the space of the object. A coordinate system has to be introduced for the mathematical expressions of physical quantities. The Cartesian coordinate is an universal choice, and especially suitable for numerical calculations. In an n-dimensional space, the position vector is $\mathbf{r} = (x_1, x_2, ..., x_n)$ and an arbitrary vector has the form $\mathbf{A} = (A_1, A_2, ..., A_n)$, which can be easily stored in an n-dimensional array in computer. It just has to be remembered, in the numerical calculation, a step size has to be introduced for a component x_i; otherwise the memory of the position vector will be infinite. Actually this discretization is a basic problem in micromagnetics, which will be further discussed in Chap. 4.

In micromagnetics, the magnetization $\mathbf{M}$ takes the role of position vector in mechanics. The polar coordinate expression $\mathbf{M} = (M, \theta, \phi)$ of magnetization is often used, where the inclination angle $0 \leq \theta \leq \pi$ and the azimuth angle $0 \leq \phi < 2\pi$. The real polycrystalline material is complicated, usually including multilevel of microstructure and different orientation of crystal, therefore the coordinate transformation between Cartesian and polar coordinate of the same vector is important:

$$M_1 = M \sin \theta \cos \phi, \tag{2.1}$$
$$M_2 = M \sin \theta \sin \phi,$$
$$M_3 = M \cos \theta.$$

$$M = \sqrt{M_1^2 + M_2^2 + M_3^2}, \tag{2.2}$$
$$\theta = \cos^{-1}(M_3/M),$$
$$\phi = \begin{cases} \cos^{-1}\left(M_1/\sqrt{M_1^2 + M_2^2}\right) & (M_2 > 0) \\ -\cos^{-1}\left(M_1/\sqrt{M_1^2 + M_2^2}\right) & (M_2 < 0). \end{cases}$$

The accurate definitions of scalar, vector and tensor depend on the characteristics of the quantity under the rotation of coordinate system. The scalar is invariant under rotation; the vector is transformed by the rotation matrix under rotation; and the tensor is transformed by two or more rotation matrices, depending on its order.

There are two rotation matrix $\tilde{R}$ and $\tilde{R}_{\mathrm{E}}$ introduced here: $\tilde{R}$ is related to two angles θ and ϕ, which is not universal but enough in most of cases; and $\tilde{R}_{\mathrm{E}}$ is general and related to three Euler angles α, β and γ, as defined in Fig. 2.1. The rotational property of a vector $\mathbf{v}$ can be described by two angles: first the coordinates (x, y, z) rotates around y-axis by angle θ, then rotates around the old z-axis by angle ϕ, as seen in Fig. 2.1a. The unit vectors $\hat{e}'_\alpha$ of the new coordinates are transformed from $\hat{e}_\alpha$ of the

Fig. 2.1 Rotation of Cartesian coordinates. **a** Two angles, no self rotation; **b** Euler angles

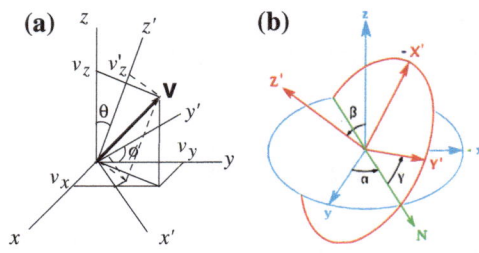

old coordinates as $\hat{e}'_\alpha = \tilde{R} \cdot \hat{e}_\alpha$ by a rotation matrix $\tilde{R}$:

$$
\tilde{R} = \begin{pmatrix} \cos\phi & -\sin\phi & 0 \\ \sin\phi & \cos\phi & 0 \\ 0 & 0 & 1 \end{pmatrix} \begin{pmatrix} \cos\theta & 0 & \sin\theta \\ 0 & 1 & 0 \\ -\sin\theta & 0 & \cos\theta \end{pmatrix}
$$

$$
= \begin{pmatrix} \cos\theta\cos\phi & -\sin\phi & \sin\theta\cos\phi \\ \cos\theta\sin\phi & \cos\phi & \sin\theta\sin\phi \\ -\sin\theta & 0 & \cos\theta \end{pmatrix} \tag{2.3}
$$

The components (v'_1, v'_2, v'_3) of a fixed vector **v** in the new coordinate system is related to (v_1, v_2, v_3) in the old coordinate system by the inverse rotation matrix or transpose of rotation matrix $\mathbf{R}^\mathrm{T}$, in a passive ("pose") manner:

$$
v'_i = R^\mathrm{T}_{ij} v_j = R_{ji} v_j \quad (i = 1, 2, 3; \ j = 1, 2, 3), \tag{2.4}
$$

$$
v_i = R_{ij} v'_j \quad\quad\quad (i = 1, 2, 3; \ j = 1, 2, 3), \tag{2.5}
$$

where the Einstein notation (auto sum over dummy indices) is used. Equations (2.4) and (2.5) can be viewed as the accurate definition of a vector.

When the rotation matrix $\tilde{R}_\mathrm{E}$ of Euler angles is utilized to define vectors, the transform between the new components (v'_1, v'_2, v'_3) and old components (v_1, v_2, v_3) is in an active ("displacement") manner, with rotations α (around z), β (around N or y' with $\gamma = 0$) and γ (around z') successively:

$$
\mathbf{v}' = \tilde{R}_\mathrm{E} \cdot \mathbf{v}, \quad v'_i = R^{ij}_\mathrm{E} v_j \tag{2.6}
$$

where the transpose of $\tilde{R}_\mathrm{E}$ related to the definition of Euler angles in Fig. 2.1b [this definition is different from the most common definition of Euler angles with a transform $x \leftrightarrow y$ for compatibility with Eq. (2.3)] is defined as:

$$\tilde{R}_{\mathrm{E}}^{\mathrm{T}} = \begin{pmatrix} c_\alpha & -s_\alpha & 0 \\ s_\alpha & c_\alpha & 0 \\ 0 & 0 & 1 \end{pmatrix} \begin{pmatrix} c_\beta & 0 & s_\beta \\ 0 & 1 & 0 \\ -s_\beta & 0 & c_\beta \end{pmatrix} \begin{pmatrix} c_\gamma & -s_\gamma & 0 \\ s_\gamma & c_\gamma & 0 \\ 0 & 0 & 1 \end{pmatrix}$$

$$= \begin{pmatrix} c_\alpha c_\beta c_\gamma - s_\alpha s_\gamma & -c_\alpha c_\beta s_\gamma - s_\alpha c_\gamma & c_\alpha s_\beta \\ s_\alpha c_\beta c_\gamma + c_\alpha s_\gamma & -s_\alpha c_\beta s_\gamma + c_\alpha c_\gamma & s_\alpha s_\beta \\ -s_\beta c_\gamma & s_\beta s_\gamma & c_\beta \end{pmatrix} \tag{2.7}$$

where c stands for cosine and s stands for sine functions respectively. Equation (2.7) is equivalent to Eq. (2.3) if we let $\alpha = \phi$, $\beta = \theta$ and $\gamma = 0$, as stated in Fig. 2.1.

There are three kinds of operations between two vectors: dot product, cross product and diad, where the results are scalar, vector and matrix, respectively:

$$\mathbf{A} \cdot \mathbf{B} = A_i B_i = A_1 B_1 + A_2 B_2 + A_3 B_3; \tag{2.8}$$

$$\mathbf{A} \times \mathbf{B} = \begin{vmatrix} \hat{e}_1 & \hat{e}_2 & \hat{e}_3 \\ A_1 & A_2 & A_3 \\ B_1 & B_2 & B_3 \end{vmatrix} = \hat{e}_i \varepsilon_{ijk} A_j B_k; \tag{2.9}$$

$$\left(\mathbf{A}\mathbf{B}^{\mathrm{T}}\right)_{ij} = A_i B_j, \tag{2.10}$$

where $\hat{e}_1, \hat{e}_2, \hat{e}_3$ are the unit vectors along x, y, z axes in Cartesian coordinates, respectively. The ε_{ijk} is a third-order antisymmetric tensor with only 6 nonzero elements $\varepsilon_{123} = \varepsilon_{231} = \varepsilon_{312} = 1 = -\varepsilon_{321} = -\varepsilon_{213} = -\varepsilon_{132}$.

The operation related to scalar is easy, just a simple multiplication. For the tensors, there are an operation called contraction: the contraction between the Nth-order tensor and the Mth-order tensor is a $|N-M|$th-order tensor:

$$X_{i_1 i_2 \ldots i_N} Y_{i_1 i_2 \ldots i_M} = Z_{i_{M+1} i_{M+2} \ldots i_N} \quad (N \geq M). \tag{2.11}$$

Therefore, the sum over dummy indices using Einstein notations is a kind of contraction, as seen in Eqs. (2.8) and (2.9). The trace of a matrix $\tilde{X}$ is a contraction between $\tilde{X}$ and the unit matrix or Kronical-delta function:

$$X_{i_1 i_2} \delta_{i_1 i_2} = X_{i_1 i_1}, \tag{2.12}$$

where the zeroth-order tensor is just a scalar; and X_{ii} is the trace of $\tilde{X}$.

As we stated before, the rotational characteristics of vector and tensor are different. If the passive and active manner of the rotation defined in Eq. (2.3) of angles θ and ϕ and Eq. (2.7) of Euler angles are used respectively, after rotation, the diad $\tilde{X} = \mathbf{A}\mathbf{B}^{\mathrm{T}}$ has the same characteristics as a tensor:

$$\tilde{X}' = (\tilde{R}^{\mathrm{T}}\mathbf{A})(\tilde{R}^{\mathrm{T}}\mathbf{B})^{\mathrm{T}} = \tilde{R}^{\mathrm{T}} \tilde{X} \tilde{R}$$
$$\tilde{X}' = (\tilde{R}_{\mathrm{E}}\mathbf{A})(\tilde{R}_{\mathrm{E}}\mathbf{B})^{\mathrm{T}} = \tilde{R}_{\mathrm{E}} \tilde{X} \tilde{R}_{\mathrm{E}}^{\mathrm{T}}. \tag{2.13}$$

These characteristics under rotation is very useful when we calculate the demagnetizing matrices of micromagnetic cells in a thin film or a device.

The Maxwell equations include the differentiation of electric and magnetic vector fields $\mathbf{E}$ and $\mathbf{B}$ in three-dimensional (3-D) space. When the vector analysis is applied to calculus in space, the differentiation in one-dimension becomes the gradient operator in three-dimensions. The express of the gradient operator ∇ and the Laplace operator ∇^2 depends on the coordinate system. In Cartesian coordinates $\hat{e}_1, \hat{e}_2, \hat{e}_3$ are constant vectors, but in cylindrical coordinates two out of three local unit vectors $\hat{e}_\rho, \hat{e}_\phi$ vary with position $\mathbf{r}$, and in spherical coordinates all three local unit vectors $\hat{e}_r, \hat{e}_\theta, \hat{e}_\phi$ vary with $\mathbf{r}$. Thus when the ∇ is acting on a vector $\mathbf{A} = \hat{e}_i A_i$ it is much more complicated in cylindrical and spherical coordinates.

In this book, we will just use Cartesian coordinates to build up the regular mesh of micromagnetic cells, therefore things become much easier. When the gradient operator acts on a vector, there are also three kinds of differentiation operations: dot product (divergence), cross product (curl) and diad:

$$\nabla = \hat{e}_i \partial_i = \hat{e}_x \frac{\partial}{\partial x} + \hat{e}_y \frac{\partial}{\partial y} + \hat{e}_z \frac{\partial}{\partial z} \tag{2.14}$$

$$\nabla \cdot \mathbf{A} = \partial_i A_i = \frac{\partial A_x}{\partial x} + \frac{\partial A_y}{\partial y} + \frac{\partial A_z}{\partial z}, \tag{2.15}$$

$$\nabla \times \mathbf{A} = \hat{e}_i \varepsilon_{ijk} \partial_j A_k, \tag{2.16}$$

$$\left(\nabla \mathbf{A}^{\mathrm{T}} \right)_{ij} = \partial_i A_j. \tag{2.17}$$

The operations between two ∇ operators (if we let $\mathbf{A} = \nabla$) can also be performed. The dot product of two ∇ is the famous Laplacian ∇^2, which is a scalar operator; the cross product of two ∇ is zero when it acts on a scalar field; and the diad of two ∇ often appears in the calculation of demagnetizing matrix.

2.1.2 Maxwell's Equations

The development of electromagnetism in the eighteenth and nineteenth centuries has been reviewed in Sect. 1.1. In 1861–1862, inspired by Faraday's thought of electromagnetic induction in dielectrics, Maxwell brought up the concept of "displacement current" in his paper "On Physical Lines of Force", and he added this extra term of effective current in the Ampere's law. In his book *A Treatise on Electricity and Magnetism* finally published in 1873, the "general equations of electromagnetic field" were derived and discussed in electric and magnetic media, and the system of units were close to the cgs units, except that the speed of light c did not appear as in the modern form of the Maxwell's equations:

Fig. 2.2 Definition of magnetic induction **B** or magnetic field **H** by Maxwell: the force acting on an unit dipole at the center of cavity if $b \ll a$ or $b \gg a$, respectively

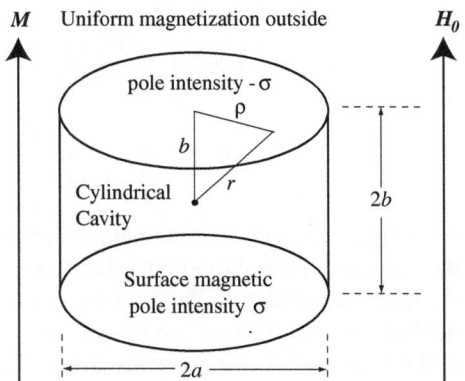

$$\nabla \cdot \mathbf{D} = 4\pi \rho_0, \tag{2.18}$$

$$\nabla \cdot \mathbf{B} = 0, \tag{2.19}$$

$$\nabla \times \mathbf{E} = -\frac{1}{c}\frac{\partial \mathbf{B}}{\partial t}, \tag{2.20}$$

$$\nabla \times \mathbf{H} = \frac{4\pi}{c}\mathbf{j}_0 + \frac{1}{c}\frac{\partial \mathbf{D}}{\partial t}, \tag{2.21}$$

where ρ_0 and $\mathbf{j}_0$ are the charge density and current density of free charge. The Coulomb's law in Eq. (2.18) was given in Sect. 77 of Maxwell's book; the Faraday's law of electromagnetic induction in Eq. (2.20) was derived in Sect. 598 by introducing the magnetic vector potential; the Gauss' law for magnetism in Eq. (2.19) was stated in Sect. 604; and the Ampere's law with the modification of displacement current in Eq. (2.21) was finally discussed in Sects. 607–610 [3].

There are four electromagnetic field in the Maxwell's equations in media: the electric displacement **D**, electric field **E**, magnetic induction **B** and magnetic field **H**; therefore the Maxwell's equations have to be solved self-consistently with the polarization equation of dielectrics in Eq. (1.4), the magnetization equation of magnetic media in Eq. (1.5), and the Ohm's law of conductors in Eq. (1.6).

The relationships among **B**, **H** and magnetization **M** of magnetic materials are actually the central topic of this book. When Maxwell introduced the magnetization equation of magnetic media in Sects. 395–400 [3], the magnetic induction **B** and magnetic field **H** were defined following Fig. 2.2. If there is a cylindrical cavity in an uniform magnetic medium with magnetization **M**, and the axis of the cylinder is parallel to **M**, the magnetic induction **B** is the force acting on the unit magnetic dipole at the center when the cavity is extremely flat ($b \ll a$), and the magnetic field **H** is the force on unit dipole when the cavity is extremely thin and long ($b \gg a$). The effective magnetic field (Maxwell called it "force" [3]) acting on the unit magnetic dipole can be found by a simple integration, by considering the contributions of the effective surface magnetic pole intensity $\pm\sigma = \pm M$ on the bottom/top surface:

$$F_z = H_0 + 2 \int\limits_0^a \frac{\sigma b \, 2\pi\rho\mathrm{d}\rho}{\left(\rho^2 + b^2\right)^{3/2}} = H_0 + 4\pi\sigma b \int\limits_b^{\sqrt{a^2+b^2}} \frac{r\mathrm{d}r}{r^3}$$

$$= H_0 + 4\pi\sigma \left(1 - \frac{b}{\sqrt{a^2+b^2}}\right), \tag{2.22}$$

where $\mathbf{H}_0$ is an uniform external magnetic field, and the "force" $\mathbf{F} = F_z\hat{e}_z$ is along the z-axis or axis of the cylinder. In a flat cavity ($b \ll a$), $\mathbf{F} = \mathbf{H}_0 + 4\pi\mathbf{M}$ just equals the magnetic induction $\mathbf{B}$; in a long cavity ($b \gg a$), $\mathbf{F} = \mathbf{H}_0$ equals the magnetic field $\mathbf{H}$, because there is no other terms such as the exchange field acting on the unit dipole. These definitions of $\mathbf{B}$ and $\mathbf{H}$ are related to the demagnetizing field in magnetic media, which will be further discussed in the next section.

The experimental verifications of Maxwell's equations have been performed at a large scale, from microscopic to macroscopic. The electromagnetic field of fundamental particles in vacuum is micro-electric field $\mathbf{e}$ and micro-magnetic field $\mathbf{h}$. The electric field $\mathbf{E}$ and magnetic induction $\mathbf{B}$ in Maxwell's equations in media from Eq. (2.18) to Eq. (2.21) are actually the statistical average of the microscopic electromagnetic field $\mathbf{e}$ and $\mathbf{h}$ in an element of media:

$$\mathbf{E} = \langle\mathbf{e}\rangle_{\text{element}}, \quad \mathbf{B} = \langle\mathbf{h}\rangle_{\text{element}}. \tag{2.23}$$

The average is necessary because, microscopically, the atoms are vibrating, the electrons are moving; therefore the micro-electric field $\mathbf{e}$ and micro-magnetic field $\mathbf{h}$ are highly nonuniform in time and space. In the studies of macroscopic electric or magnetic materials, these fluctuations of microscopic electromagnetic field around fundamental particles need not to be considered, thus an average in an element, at least with a size of several conventional unit cells, can be made to obtain the macroscopic $\mathbf{E}$ and $\mathbf{B}$. However, if the scattering of external particle or wave with the matter has to be considered, quantum excitation, absorbing or emission processes are involved, the average in Eq. (2.23) is no longer appropriate.

The Maxwell's equations in vacuum have an beautiful symmetric form between electric and magnetic phenomena, except that there is no intrinsic "magnetic charge" for fundamental particles:

$$\nabla \cdot \mathbf{e} = 4\pi\rho, \tag{2.24}$$

$$\nabla \cdot \mathbf{h} = 0, \tag{2.25}$$

$$\nabla \times \mathbf{e} = -\frac{1}{c}\frac{\partial \mathbf{h}}{\partial t}, \tag{2.26}$$

$$\nabla \times \mathbf{h} = \frac{4\pi}{c}\mathbf{j} + \frac{1}{c}\frac{\partial \mathbf{e}}{\partial t}. \tag{2.27}$$

The concepts of polarization $\mathbf{P}$ and magnetization $\mathbf{M}$ do not exist microscopically, because the microscopic view is a view of elementary particles in vacuum. Therefore Eqs. (2.24)–(2.27) are enough to solve the microscopic electromagnetic field if the sources $(\rho, \mathbf{j})$ of particles are known. In early twentieth century, Dutch physicist Hendrik Antoon Lorentz pointed out that, when the interactions among the electromagnetic field and fundamental particles are considered, there must be a formula for electromagnetic force to make the problem complete. This is the famous Lorentz force for a fundamental particle with charge q and velocity $\mathbf{v}$:

$$\mathbf{F} = q\mathbf{e} + \frac{q}{c}\mathbf{v} \times \mathbf{h} \qquad (2.28)$$

Einstein called the four equations from Eq. (2.24) to Eq. (2.27), plus the Eq. (2.28), a complete set of Maxwell–Lorentz theory of electromagnetism.

It would be important to state the Maxwell's equations in the SI or MKS units, especially for the macroscopic problems in media. The unit transformation has been introduced in Tables 1.1 and 1.2 in Sect. 1.1. Actually we can follow some simple rules to transform the Maxwell's equations in cgs units from Eq. (2.18) to Eq. (2.21) into the MKS units (in cgs units, $\mathbf{E}$, $\mathbf{D}$, $\mathbf{H}$ and $\mathbf{B}$ have the same dimension; but in MKS units, both the ratio $[E/B]$ and $[H/D]$ have a dimension of speed c, the unit of $\mathbf{D}$ is C/m^2, and the unit of $\mathbf{H}$ is A/m):

$$\nabla \cdot \mathbf{D} = \rho_0, \qquad (2.29)$$

$$\nabla \cdot \mathbf{B} = 0, \qquad (2.30)$$

$$\nabla \times \mathbf{E} = -\frac{\partial \mathbf{B}}{\partial t}, \qquad (2.31)$$

$$\nabla \times \mathbf{H} = \mathbf{j}_0 + \frac{\partial \mathbf{D}}{\partial t}, \qquad (2.32)$$

where $(\rho_0, \mathbf{j}_0)$ are source of "free charge" in conductors. The Maxwell's equations Eqs. (2.29)–(2.32) also have to be solved self-consistently with the polarization equation of electric media in Eq. (1.4), the magnetization equation of magnetic media in Eq. (1.5), and the Ohm's law of conductors in Eq. (1.6).

2.2 Green's Function and Demagnetizing Matrix

The Green's function was brought up by George Green in 1824 to solve the electrostatic problem. If a point charge q is put at point $\mathbf{r}_0$ in a space with a conductor, the potential at another point $\mathbf{r}$ is contributed by the Coulomb potential from q and the potential from the induced charge on the surface of the conductor:

$$V(\mathbf{r}) = \frac{q}{|\mathbf{r} - \mathbf{r}_0|} + \iint d^2\mathbf{r}' G(\mathbf{r}, \mathbf{r}')\sigma(\mathbf{r}') = \frac{q}{|\mathbf{r} - \mathbf{r}_0|} + \iint d^2\mathbf{r}' \frac{\sigma(\mathbf{r}')}{|\mathbf{r} - \mathbf{r}'|}, \quad (2.33)$$

where the Green's function $G(\mathbf{r}, \mathbf{r}') = 1/|\mathbf{r} - \mathbf{r}'|$ is the solution in an infinite space of the Poisson equation with an unit point charge located at $\mathbf{r}'$:

$$\nabla^2 G(\mathbf{r}, \mathbf{r}') = -4\pi \delta^3(\mathbf{r} - \mathbf{r}'). \tag{2.34}$$

The Poisson equation developed by Siméon Poisson in 1811 is consistent with differential form of the Coulomb's law in Eq. (2.24) if we define the electric field $\mathbf{e} = -\nabla V$. The Green's function $G(\mathbf{r}, \mathbf{r}') = G(\mathbf{r}', \mathbf{r}) = 1/|\mathbf{r} - \mathbf{r}'|$ in an infinite space thus has the famous $1/r$ potential form, the same as the Coulomb potential.

The Green's function is important because it keeps the same form when the external charge q and the induced surface charge σ vary with time. When we try to solve the magnetic properties of a thin film or a device, although there is no intrinsic magnetic charge, the induced magnetic pole density ρ_M varies with the external field $\mathbf{H}_{ext}$. This can be clarified by the differential form of the Gauss' law for magnetism in Eq. (2.19) and the magnetization equation in Eq. (1.5):

$$\nabla \cdot \mathbf{H} = 4\pi \rho_M \quad \text{(cgs)}; \qquad \rho_M = -\nabla \cdot \mathbf{M}. \tag{2.35}$$

If the magnetization $\mathbf{M}$ is uniform in a grain or cell, the magnetic pole only exists on its surface. It is easy to prove by the Gauss law that the surface magnetic pole density is $\sigma_M = \hat{n} \cdot \mathbf{M}$, where $\hat{n}$ is the unit vector normal to the surface.

In micromagnetics, the magnetic material is discretized into micromagnetic cells. Inside a cell, the moment $\mu = V_c \mathbf{M}$ is assumed to rotate uniformly. Following the spirit of the Green's function, the demagnetizing matrix $\tilde{N}$ can be defined, where the magnetization $\mathbf{M}$ in a micromagnetic cell is the "source", and the magnetostatic interaction field or demagnetizing field $\mathbf{H}_d$ is the "target":

$$\mathbf{H}_d(\mathbf{r}) = \iiint d^3\mathbf{r}' \left[-\nabla' \cdot \mathbf{M}(\mathbf{r}') \right] \frac{(\mathbf{r} - \mathbf{r}')}{|\mathbf{r} - \mathbf{r}'|^3} = -\tilde{N}(\mathbf{r}, \mathbf{0}) \cdot [4\pi \mathbf{M}], \tag{2.36}$$

$$\tilde{N}(\mathbf{r}, \mathbf{0}) = -\frac{1}{4\pi} \iiint_{V_c} d^3\mathbf{r}' \, \nabla \nabla' \frac{1}{|\mathbf{r} - \mathbf{r}'|} = -\frac{1}{4\pi} \iint_{S} d^2\mathbf{r}' \frac{(\mathbf{r} - \mathbf{r}')\hat{n}'}{|\mathbf{r} - \mathbf{r}'|^3}, \tag{2.37}$$

where the 3-D integral is made over the volume of the micromagnetic cell V_c (this is related to the fact that the magnetization $\mathbf{M}(\mathbf{r}') = \mathbf{M}$ is zero outside the cell), and the 2-D integral is made over the surface S of the cell. The $(\mathbf{r} - \mathbf{r}')\hat{n}'$ is a diad of the two vectors, and $\hat{n}'$ is the local normal at $\mathbf{r}'$ pointing outside the cell.

In micromagnetics, the demagnetizing matrix $\tilde{N}(\mathbf{r}_i, \mathbf{r}_j) = \tilde{N}(\mathbf{r}, \mathbf{0})$ only depends on the relative displacement $\mathbf{r} = \mathbf{r}_i - \mathbf{r}_j$ of the ith and jth micromagnetic cells, and it is independent of the magnetization of the cells. This characteristics is especially important when the problem is solved numerically, because $\tilde{N}$ can be computed in advance, stored and used repeatedly in calculation. Due to the intrinsic characteristics of the Green's function or the demagnetizing matrix, the trace is a conserved quantity when the target vector $\mathbf{r}$ is inside or outside of the micromagnetic cell:

$$\text{Tr} \, \tilde{N}(\mathbf{r}, \mathbf{0}) = \sum_\alpha N_{\alpha\alpha}(\mathbf{r}, \mathbf{0}) = \begin{cases} 1 & \mathbf{r} \text{ inside the cell} \\ 0 & \mathbf{r} \text{ outside the cell} \end{cases} \tag{2.38}$$

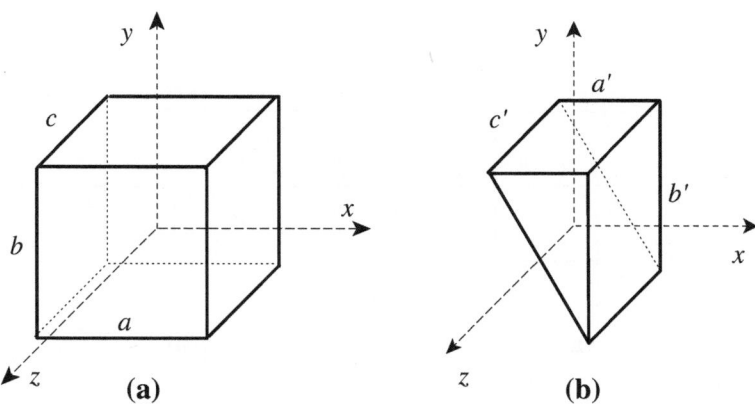

Fig. 2.3 Micromagnetic cells in a magnetic thin film. **a** Cuboid or cubic cell in the main body; **b** triangular prism cell at the edge; © [2009] IEEE. Reprinted, with permission, from Ref. [5]

Table 2.1 Demagnetizing matrix of a cuboic cell with uniform magnetization [7]

Cuboid cell size $a \times b \times c$	Integer variables $p = \pm1, q = \pm1, w = \pm1$
Intermediate variable $\mathbf{R} = (R_1, R_2, R_3)$	$R_1 = \frac{a}{2} + px \quad R_2 = \frac{b}{2} + qy, \quad R_3 = \frac{c}{2} + wz$
Demagnetizing matrix element N_{11}	$\frac{1}{4\pi} \sum_p \sum_q \sum_w \tan^{-1}[R_2 R_3/(R_1 R)]$
Demagnetizing matrix element N_{22}	$\frac{1}{4\pi} \sum_p \sum_q \sum_w \tan^{-1}[R_3 R_1/(R_2 R)]$
Demagnetizing matrix element N_{33}	$\frac{1}{4\pi} \sum_p \sum_q \sum_w \tan^{-1}[R_1 R_2/(R_3 R)]$
Demagnetizing matrix element $N_{12} = N_{21}$	$\frac{1}{8\pi} \sum_p \sum_q \sum_w pq \ln[(R - R_3)/(R + R_3)]$
Demagnetizing matrix element $N_{13} = N_{31}$	$\frac{1}{8\pi} \sum_p \sum_q \sum_w pw \ln[(R - R_2)/(R + R_2)]$
Demagnetizing matrix element $N_{23} = N_{32}$	$\frac{1}{8\pi} \sum_p \sum_q \sum_w qw \ln[(R - R_1)/(R + R_1)]$

This is an useful property to check the correctness of the demagnetizing matrix.

Modern ferromagnetic devices are mostly made by thin films, as introduced in Sect. 1.3. To discretize an arbitrary-shaped device into micromagnetic cells, there are two basic classes for the geometry of cells: cuboid or cubic cell in the main body, and triangular prism cells at the edge, as seen in Fig. 2.3. The demagnetizing matrices of either cells can be calculated by Eq. (2.37), with a sum of contributions from all surfaces of the cell.

The demagnetizing matrix of a cuboid cell with a size $a \times b \times c$ can be integrated out directly [6]. Among the nine elements, there are only two independent types:

$$\tilde{N} = -\frac{1}{4\pi} \int_{-a/2}^{a/2} \mathrm{d}x' \int_{-b/2}^{b/2} \mathrm{d}y' \int_{-c/2}^{c/2} \mathrm{d}z' \; \nabla'\nabla' \frac{1}{|\mathbf{r} - \mathbf{r}'|} \tag{2.39}$$

$$N_{11} = \frac{1}{4\pi} \sum_{p=\pm1} \iint \mathrm{d}y'\mathrm{d}z' \frac{a/2 + px}{\left[(a/2 + px)^2 + (y - y')^2 + (z - z')^2\right]^{3/2}} \tag{2.40}$$

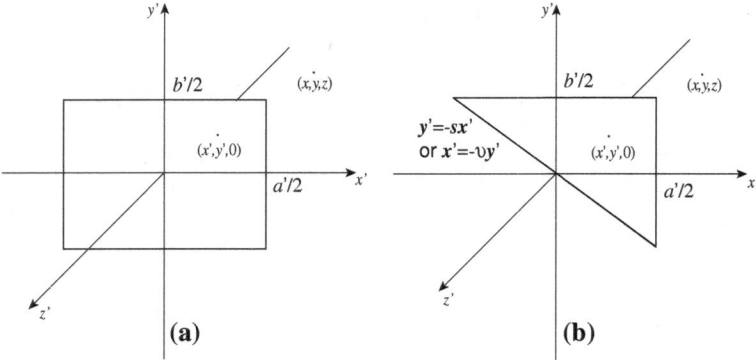

Fig. 2.4 Demagnetizing matrix contributed by poles on a 2-D surface. **a** Rectangle; **b** triangle;
© [2009] IEEE. Reprinted, with permission, from Ref. [5]

$$N_{12} = -\frac{1}{4\pi} \sum_{p=\pm 1} \sum_{q=\pm 1} pq \int \frac{dz'}{\left[(a/2 + px)^2 + (b/2 + qy)^2 + (z - z')^2\right]^{3/2}},$$
(2.41)

where the sum integer variables $p, q = \pm 1$ come from the two integral limits. The nine elements in the demagnetizing matrix of a cuboic cell are listed in Table 2.1.

The demagnetizing matrix of a cuboid cell $\tilde{N}$ or a triangular prism cell $\tilde{N}^{(t)}$ (the center of cell locates at **0**) in Fig. 2.3 can be calculated by summing over the contributions from the magnetic poles on the rectangle or triangle 2-D surfaces:

$$\tilde{N}(\mathbf{r}, \mathbf{0}) = \sum_{l=1}^{6} \tilde{R}_l \cdot \tilde{N}^{\text{rec}} \left(\tilde{R}_l^{\text{T}} \cdot (\mathbf{r} - \delta_l), \mathbf{0} \right) \cdot \tilde{R}_l^{\text{T}}$$
(2.42)

$$\tilde{N}^{(t)}(\mathbf{r}, \mathbf{0}) = \sum_{l=1}^{3} \tilde{R}_l \cdot \tilde{N}^{\text{rec}} \left(\tilde{R}_l^{\text{T}} \cdot (\mathbf{r} - \delta_l), \mathbf{0} \right) \cdot \tilde{R}_l^{\text{T}}$$
(2.43)

$$+ \sum_{n=1}^{2} \tilde{R}_n \cdot \tilde{N}^{\text{tri}} \left(\tilde{R}_n^{\text{T}} \cdot (\mathbf{r} - \delta_n), \mathbf{0} \right) \cdot \tilde{R}_n^{\text{T}}$$

where $\tilde{N}^{\text{rec}}$ or $\tilde{N}^{\text{tri}}$ is the demagnetizing matrix of a rectangular or a right-angle triangular surface located in a fixed plane (say $z' = 0$), as seen in Fig. 2.4; $\tilde{R}$ is the 3-D rotational matrix between the real position of the respective surface in a micromagnetic cell ($x = \pm a/2, y = \pm b/2, z = \pm c/2$ for a cubic cell; or $x = a'/2$, $y = b'/2, y = -sx, z = \pm c'/2$ for a triangular prism cell) and the suppositional surface in the fixed $z' = 0$ plane; and δ is the displacement vector from the center of a cell to the face center of a square/rectangular surface, or from the center to the midpoint of the hypotenuse of a triangular surface, as labeled in Fig. 2.4.

Table 2.2 Demagnetizing matrix of a $a' \times b'$ rectangular surface at $z' = 0$ with uniform pole

Integer variables $q = \pm 1$, $w = \pm 1$, Intermediate variable $\mathbf{R} = (R_1, R_2, R_3)$
$R_1 = \frac{a'}{2} + qx$, $R_2 = \frac{b'}{2} + wy$, $R_3 = z$, $R = |\mathbf{R}| = \sqrt{R_1^2 + R_2^2 + R_3^2}$
$N_{13}^{\text{rec}} = -\frac{1}{4\pi} \sum_q \sum_w qw \ln(R - wR_2) = -\frac{1}{8\pi} \sum_q \sum_w q \ln[(R - R_2)/(R + R_2)]$
$N_{23}^{\text{rec}} = -\frac{1}{4\pi} \sum_q \sum_w qw \ln(R - qR_1) = -\frac{1}{8\pi} \sum_q \sum_w w \ln[(R - R_1)/(R + R_1)]$
$N_{33}^{\text{rec}} = -\frac{1}{4\pi} \sum_q \sum_w \arctan[R_1 R_2/(R_3 R)]$

Table 2.3 Demagnetizing matrix of a $a' \times b'$ right-angle triangular surface at $z' = 0$ with hypotenuse $y' = -sx'$ or $x' = -vy'$

$R_{\text{I}} = \sqrt{\left(\frac{a'}{2} - x\right)^2 + R_2^2 + R_3^2}$, $R_{\text{II}} = \sqrt{R_1^2 + \left(\frac{b'}{2} - y\right)^2 + R_3^2}$
$c_1 = \frac{y - vx}{1 + v^2}$, $c_2^2 = \frac{r^2}{1 + v^2} - c_1^2$, $P_2 = \frac{b'}{2} + wc_1$, $P = \sqrt{P_2^2 + c_2^2}$
$c_3 = \frac{x - sy}{1 + s^2}$, $c_4^2 = \frac{r^2}{1 + s^2} - c_3^2$, $Q_1 = \frac{a'}{2} + qc_3$, $Q = \sqrt{Q_1^2 + c_4^2}$
$c_5 = y - iz$, $\sqrt{(c_1 - c_5)^2 + c_2^2} = Ae^{i\theta'/2}$
$V_\pm = -wP_2 + \sqrt{P_2^2 + c_2^2} + c_1 - c_5 \pm Ae^{i\theta'/2} = |V_\pm|\, e^{i\phi_\pm}$,
$N_{13}^{\text{tri}} = -\frac{1}{4\pi} \sum_w w \left\{ \frac{1}{\sqrt{1+v^2}} \ln(P - wP_2) - \ln(R_{\text{I}} - wR_2) \right\}$
$N_{23}^{\text{tri}} = -\frac{1}{4\pi} \sum_q q \left\{ \frac{1}{\sqrt{1+s^2}} \ln(Q - qQ_1) - \ln(R_{\text{II}} - qR_1) \right\}$
$N_{33}^{\text{tri}} = -\frac{1}{4\pi} \sum_w \arctan[(a'/2 - x)R_2/(R_3 R_{\text{I}})]$
$\qquad -\frac{1}{4\pi} \frac{zv}{\sqrt{1+v^2}} \sum_w \frac{w}{A} \left[+ \ln \left|\frac{V_+}{V_-}\right| \cos\frac{\theta'}{2} + (\phi_+ - \phi_-) \sin\frac{\theta'}{2} \right]$
$\qquad +\frac{1}{4\pi} \frac{x+vy}{\sqrt{1+v^2}} \sum_w \frac{w}{A} \left[- \ln \left|\frac{V_+}{V_-}\right| \sin\frac{\theta'}{2} + (\phi_+ - \phi_-) \cos\frac{\theta'}{2} \right]$

Definitions of R_1, R_2, R_3, q and w are the same as in Table 2.2

The demagnetizing matrix $\tilde{N}^{\text{rec}}(\mathbf{r}, \mathbf{0})$ or $\tilde{N}^{\text{tri}}(\mathbf{r}, \mathbf{0})$, contributed by magnetic poles on a rectangular or a triangular surface, can be calculated by the 2-D integral in Eq. (2.37), where the target position locates at $\mathbf{r} = (x, y, z) = (r_1, r_2, r_3)$, as seen in Fig. 2.4. The integrations for nonzero elements of $\tilde{N}^{\text{rec}}(\mathbf{r}, \mathbf{0})$ and $\tilde{N}^{\text{tri}}(\mathbf{r}, \mathbf{0})$ [5]

$$N_{\alpha 3}^{\text{rec}}(\mathbf{r}, \mathbf{0}) = -\frac{1}{4\pi} \int_{-a'/2}^{a'/2} dr_1' \int_{-b'/2}^{b'/2} dr_2' \frac{r_\alpha - r_\alpha'}{[(r_1 - r_1')^2 + (r_2 - r_2')^2 + r_3^2]^{3/2}} \quad (2.44)$$

$$N_{\alpha 3}^{\text{tri}}(\mathbf{r}, \mathbf{0}) = -\frac{1}{4\pi} \int_{-b'/2}^{b'/2} dy' \int_{-vy'}^{a'/2} dx' \frac{r_\alpha - r_\alpha'}{[(r_1 - r_1')^2 + (r_2 - r_2')^2 + r_3^2]^{3/2}} \quad (2.45)$$

$$= -\frac{1}{4\pi} \int_{-a'/2}^{a'/2} dx' \int_{-sx'}^{b'/2} dy' \frac{r_\alpha - r_\alpha'}{[(r_1 - r_1')^2 + (r_2 - r_2')^2 + r_3^2]^{3/2}} \quad (2.46)$$

Fig. 2.5 Bloch wall
between two domains:
$m_z = \tanh(x/a)$ where wall
width $a = \sqrt{A^*/K_1}$

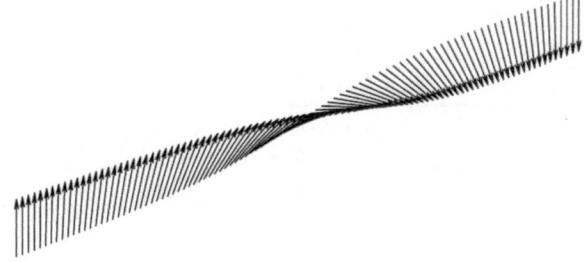

are given in the Appendix. The analytical expressions of $N_{\alpha 3}^{\mathrm{rec}}(\mathbf{r}, \mathbf{0})$ and $N_{\alpha 3}^{\mathrm{tri}}(\mathbf{r}, \mathbf{0})$
($\alpha = 1, 2, 3$) are given in Tables 2.2 and 2.3, respectively.

Using the results of the surface demagnetizing matrices given in Tables 2.2 and
2.3, together with the definitions of the demagnetizing matrices of a cubic/cuboid cell
and a triangular prism cell in Eqs. (2.42) and (2.43) respectively, the magnetostatic
interaction field can be calculated for any micromagnetic cell in a ferromagnetic thin
film or device with a flat structure. In more complicated device with 3-D structures,
the surface of any edge micromagnetic cell can still be divided into rectangles and
right-angle triangles, thus the respective demagnetizing matrix can also be found
analytically using the similar methods as in Eq. (2.43).

2.3 Landau–Lifshitz Equations

In 1935, Landau and Lifshitz brought up a free energy expression for ferromagnetic
materials [8]; furthermore, a dynamic equation of spins, called LL equation later,
was given to analyze the motion of Bloch domain wall in Fig. 2.5. These works built
the fundamentals of applied magnetic theory.

As discussed in Sect. 1.2, in 1907, Pierre-Ernest Weiss brought up the concept
of a huge spontaneous magnetic field to explain the alignment of the elementary
magnets (atomic spins) in a ferromagnetic material even without an external field.
The spontaneous magnetization is related to the crystal structure, which orients to
the easy axis. A single crystal contains a large amount of "domains", and the size of
a Weiss domain is assumed to be 10–100 nm.

The modern magnetization-curve theory was founded by Akulov and Becker.
In 1929, Akulov derived the anisotropy energy for Fe and Ni [9]; in 1930, Becker
introduced a rigid rotational model to find the loops of a particle [10]. In 1932,
Bloch worked out a structure of the boundary wall between two Weiss domains [11],
now known as the "Bloch wall", although its common expression $m_z = \tanh(x/a)$
was given by Landau [8]. The concept of "Micromagnetics" was brought up by
Brown in 1958. The rise of computational science in 1940s made it possible to
solve the LL equations numerically, which led to the micromagnetic theory today
for ferromagnetic materials.

2.3.1 Free Energy and Effective Field

Landau was a master of phenomenological theory in condensed matter physics. In the free energy $\mathscr{F}$ of a ferromagnet, the magnetization vector $\mathbf{M} = M_s \hat{m}$ takes the role of "position vector" $\mathbf{r}$ as in mechanics, the Zeeman energy $\mathscr{E}_{ext}$ due to the external magnetic field, the anisotropy energy $\mathscr{E}_a$ describing the Weiss spontaneous magnetization, the exchange energy $\mathscr{E}_{ex}$ of Heiseberg model, the magnetostatic interaction energy $\mathscr{E}_m$, and the magneto-elastic energy $\mathscr{E}_\sigma$ are included [1]:

$$\mathscr{F}(\{\hat{m}\}) = \mathscr{F}_0 + \mathscr{E}_{ext} + \mathscr{E}_a + \mathscr{E}_{ex} + \mathscr{E}_m + \mathscr{E}_\sigma, \tag{2.47}$$

$$\mathscr{E}_{ext} = - \iiint d^3\mathbf{r} \ M_s \hat{m}(\mathbf{r}) \cdot \mathbf{H}_{ext}(\mathbf{r}), \tag{2.48}$$

$$\mathscr{E}_a = \iiint d^3\mathbf{r} \ \left(K_{ij}^{(1)} m_i m_j + K_{ijkl}^{(2)} m_i m_j m_k m_l + \mathrm{o}(m^6) \right), \tag{2.49}$$

$$\mathscr{E}_{ex} = \frac{1}{2} \iiint d^3\mathbf{r} \ (2A_{ij}^*)(\partial_i m_l)(\partial_j m_l), \tag{2.50}$$

$$\mathscr{E}_m = -\frac{1}{2} \iiint d^3\mathbf{r} \ M_s \hat{m}(\mathbf{r}) \cdot \mathbf{H}_d(\mathbf{r})$$
$$= \frac{1}{2}(4\pi M_s^2) \sum_{\mathbf{r}} V_{\mathbf{r}} \sum_{\mathbf{r}'} \hat{m}(\mathbf{r}) \cdot \tilde{N}(\mathbf{r}, \mathbf{r}') \cdot \hat{m}(\mathbf{r}'), \tag{2.51}$$

$$\mathscr{E}_\sigma = - \iiint d^3\mathbf{r} \ a_{ijkl} \ \sigma_{ij} \ m_k m_l, \tag{2.52}$$

where Einstein notation is used to sum all the dummy indices from 1 to 3. In Eq. (2.49), the first two orders of crystalline anisotropy energy are included, which are both important in crystals with different symmetries. In Eq. (2.50), the exchange constant $2A^*$ is the parameter α in Landau and Lifshitz's original paper [8]; $2A^*$ but not A^* is utilized because in this way A^* has a direct relationship with the exchange energy J_e in the Heisenberg model discussed in Sect. 1.2: $A^* = J_e/R$ where R is the nearest neighbor (NN) atomic distance. In Eq. (2.51), the demagnetizing field $\mathbf{H}_d$ is similar to the one given in Eq. (2.36), except that the $\mathbf{H}_d$ here is contributed by all micromagnetic cells $\mathbf{r}'$ in the medium; in the second line of this equation, a sum but not an integral over the volume $V_{\mathbf{r}}$ is utilized, because the concept of "demagnetizing matrix" $\tilde{N}$ in Eq. (2.37) has to be defined in a finite-size volume $V_{\mathbf{r}'}$. In Eq. (2.52), σ_{ij} is the stress matrix, and the fourth-order tensor parameter a_{ijkl} links the magnetic property and the mechanical property in a crystal; this term can also be used at the interface of thin films, where the stress σ_{ij} is nonzero only in several atomic planes, but this will be very important to explain the thin film properties.

The free energy of crystalline anisotropy in Eq. (2.49) needs to be further discussed for different crystal symmetries. In general, we can prove the characteristics of the parameter matrix $K_{ij}^{(1)}$ by the rotational invariant under all rotational matrices $\tilde{R}$ of a certain symmetry:

$$\tilde{K}^{(1)} = \tilde{R} \cdot \tilde{K}^{(1)} \cdot \tilde{R}^{\mathrm{T}} \tag{2.53}$$

For cubic crystals such as iron and nickel, if we use two successive $\tilde{R}$ of C_4 operation, the off-diagonal matrix elements $K_{ij}^{(1)}$ ($i \neq j$) can be proved to be zero; then by using $\tilde{R}$ of C_3 operation, we can proved that $K_{11}^{(1)} = K_{22}^{(1)} = K_{33}^{(1)}$. However, under the constraint of $\hat{m}^2 = 1$, the term of $K_{ij}^{(1)}$ is trivial. The next order parameter $K_{ijkl}^{(2)}$ can be treated as the diad of two matrices. Still, by two successive $\tilde{R}$ of C_4 operation and another $\tilde{R}$ of C_3 operation, we can prove that there are only two kinds of nonzero independent parameters in this term: $K_a = K_{iijj}^{(2)} = K_{ijij}^{(2)} = K_{ijji}^{(2)}$ and $K_b = K_{iiii}^{(2)}$, and the cubic crystalline anisotropy energy is in the form [9]:

$$\mathscr{E}_a^{(c)} = \iiint d^3\mathbf{r} \; \Big\{ 2K_1[(\hat{m} \cdot \hat{k}_1)^2(\hat{m} \cdot \hat{k}_2)^2 + (\hat{m} \cdot \hat{k}_2)^2(\hat{m} \cdot \hat{k}_3)^2$$
$$+ (\hat{m} \cdot \hat{k}_3)^2(\hat{m} \cdot \hat{k}_1)^2] + K_2(\hat{m} \cdot \hat{k}_1)^2(\hat{m} \cdot \hat{k}_2)^2(\hat{m} \cdot \hat{k}_3)^2 \Big\} \quad (2.54)$$

The $\hat{k}_1, \hat{k}_2, \hat{k}_3$ are cubic axes; $K_1 = 3K_a - K_b$ is positive for Fe and negative for Ni.

For a hexagonal crystal such as cobalt, it is easy to prove that, for $K_{ij}^{(1)}$ matrix, there are only two nonzero independent elements $K_{11}^{(1)} = K_{22}^{(1)}$ and $K_{33}^{(1)}$ if we use $\tilde{R}$ of C_6 group with $\theta = \pi$ and $\theta = \pi/3$. Furthermore, by using the constraint $m_x^2 + m_y^2 + m_z^2 = 1$, there is only one independent parameter $K_1 = K_{11}^{(1)} - K_{33}^{(1)}$ left. Then, the hexagonal anisotropy energy must be in the form [1]:

$$\mathscr{E}_a^{(h)} = \iiint d^3\mathbf{r} \; \Big\{ K_1[\hat{m} \times \hat{k}_c]^2 + K_2[\hat{m} \times \hat{k}_c]^4 \Big\} \quad (2.55)$$

where K_2 is the second-order term which can be proved by the symmetry of $K_{ijkl}^{(2)}$.

The magnetic recording media such as CoPt and FePt have a tetragonal symmetry of $L1_0$ phase. The fourfold symmetry axis is the c-axis or z-axis, by using two successive $\tilde{R}$ of C_4 operation around c-axis, we can prove that $K_{u1} = K_{11}^{(1)} = K_{22}^{(1)}$. Both the hexagonal and tetragonal anisotropy are called "uniaxial anisotropy", because the first term in their crystalline anisotropy is $K_{u1} \sin^2 \theta$ or $-K_{u1} \cos^2 \theta$ around the c-axis. In the term of $K_{ijkl}^{(2)}$, there are also two classes of independent parameters. Then the tetragonal crystalline anisotropy energy is in the form [1]:

$$\mathscr{E}_a^{(t)} = \iiint d^3\mathbf{r} \; \Big\{ -K_{u1}(\hat{m} \cdot \hat{k}_c)^2 - K_{u2}\Big[1 - (\hat{m} \cdot \hat{k}_c)^2\Big]^2$$
$$- K_c(\hat{m} \cdot \hat{k}_a)^2(\hat{m} \cdot \hat{k}_b)^2 \Big\} \quad (2.56)$$

where $\hat{k}_a$, $\hat{k}_b$, $\hat{k}_c$ are the crystal axes for a tetragonal conventional unit cell. The details of the second-order parameters K_{u2} and K_c depend on the degree of order of the alloy, and usually K_c is more important and harder to control in experiment.

Table 2.4 Anisotropy constants and exchange field constants $H_e = 2A^*/(M_s D^2)$

FM Crystals	Fe	Co	Ni
M_s(emu/g)	221.71 ± 0.08	162.55	58.57 ± 0.03
M_s(emu/cm^3)	1742.6	1446.7	521.3
K_1(erg/cm^3)	4.81×10^5	4.12×10^6	-5.5×10^4
K_2(erg/cm^3)	1.2×10^3	1.43×10^6	-2.5×10^4
Easy axis	$\hat{k}_1 , \hat{k}_2 , \hat{k}_3$	$\hat{k}_c$	$\hat{k}_1 + \hat{k}_2 + \hat{k}_3$
$H_k = 2K_1/M_s$(Oe)	552	5,696	-221
H_e(Oe) with $D = 10\,$nm	1,148	1,382	3,827
H_e(Oe) with $D = 2\,$nm	28,700	34,550	95,675

D is the micromagnetic cell size, A^* is assumed to be on the order of 1×10^{-6} erg/cm, for Fe, Co and Ni

The micromagnetics is not an atomic or electronic scale theory; therefore the basic magnetic parameters such as M_s, K_1 and A^* have to be treated as the input of the model. Actually in practical alloy or composite magnetic materials, these basic parameters can not be provided by either accurate theoretical calculation or direct experimental measurement. The experiment-simulation cycles have to be done to fit these parameters M_s, K_1 and A^*. In Table 1.3, the structure and crystalline characteristics of ferromagnetic crystals have been given, here in Table 2.4, the anisotropy energy constant and exchange field constant of Fe, Co, Ni are given respectively.

In a magnetic material, the effective magnetic field $\mathbf{H}_{\text{eff}}$ felt by a micromagnetic cell at $\mathbf{r}$ can be found by the variation $\mathbf{H}_{\text{eff}}(\mathbf{r}) = -\delta\mathscr{F}/\delta(M_s\hat{m}(\mathbf{r}))$ in the continuum integration form in Eqs. (2.49)–(2.52) or by the derivation $\mathbf{H}_{\text{eff}} = -\partial\mathscr{F}/\partial(V_{\mathbf{r}}M_s\hat{m}(\mathbf{r}))$ in the discretized summation form over $\mathbf{r}$ in Eq. (2.51) (cgs units):

$$\mathbf{H}_{\text{eff}}(\mathbf{r}) = \mathbf{H}_{\text{ext}}(\mathbf{r}) + \mathbf{H}_a(\mathbf{r}) + \mathbf{H}_{\text{ex}}(\mathbf{r}) + \mathbf{H}_m(\mathbf{r}) + \mathbf{H}_\sigma(\mathbf{r}). \tag{2.57}$$

$$\mathbf{H}_a(\mathbf{r}) = H_k \left(\hat{e}_i \, K_{ij}^{(1)} m_j + \hat{e}_i \, 2K_{ijkl}^{(2)} m_j m_k m_l + \mathrm{o}(m^5) \right) / K_1, \tag{2.58}$$

$$\mathbf{H}_{\text{ex}}(\mathbf{r}) = (2\hat{e}_l/M_s) \, A_{ij}^* \partial_i \partial_j m_l \simeq H_e \sum_{\mathbf{r}'}^{\text{NN}} \left(\hat{m}(\mathbf{r}') - \hat{m}(\mathbf{r}) \right), \tag{2.59}$$

$$\mathbf{H}_m(\mathbf{r}) = -(4\pi M_s) \sum_{\mathbf{r}'} \tilde{N}(\mathbf{r}, \mathbf{r}') \cdot \hat{m}(\mathbf{r}'), \tag{2.60}$$

$$\mathbf{H}_\sigma(\mathbf{r}) \simeq H_{\text{me}} \left(\hat{m}(\mathbf{r}) \cdot \hat{e}_x \right)\hat{e}_x + H_{\text{me}}' \left[(\hat{m}(\mathbf{r}) \cdot \hat{e}_y)\hat{e}_x + (\hat{m}(\mathbf{r}) \cdot \hat{e}_x)\hat{e}_y \right], \tag{2.61}$$

where Einstein notation is still used in three of previous equations. In Eq. (2.58), the anisotropy field constant $H_k = 2K_1/M_s$, and the relationship among K_1, $K_{ij}^{(1)}$ and $K_{ijkl}^{(2)}$ has been discussed in crystals with different symmetries. For cubic symmetry, the term of $K_{ij}^{(1)}$ will not appear, since it is a trivial or constant term.

Furthermore, if a clearer vector form of anisotropy field is to be used, the derivative $\mathbf{H}_a = -\partial \mathscr{E}_a/\partial(V_{\mathbf{r}} M_s \hat{m}(\mathbf{r}))$ can be taken directly for Eqs. (2.55)–(2.56).

In Eq. (2.59), the exchange field constant is $H_e = 2A^*/(M_s D^2)$ where D is the micromagnetic cell size. In the original form of Landau and Lifshitz [1], the exchange constant A_{jl}^* also depends on the crystal symmetry. Inside a micromagnetic cell, the magnetic moment is assumed to rotate uniformly. With two neighbor cells at a distance of D, which is usually larger than 1 nm, at least one order larger than atomic distance R, the exchange field constant $H_e = 2J_e/(M_s D^2 R)$ is much smaller than the Weiss field $H_E \simeq z J_e/(M_s R^3) \sim 10^7$ Oe among neighbor atoms, where z is the number of NN atomic spins. Therefore, we can make an approximation $A*_{ij} \simeq A^* \delta_{ij}$ and omit the crystal asymmetry in Eq. (2.59), where the sum over $\mathbf{r}'$ is only taken over the NN cells of the cell at $\mathbf{r}$.

In Eq. (2.60), the sum over $\mathbf{r}'$ is taken over micromagnetic cells discretized at a distance D in any dimension, similar to Eq. (2.59). When $\mathbf{r}' = \mathbf{r}$, the term in the sum $-B_s \tilde{N}(\mathbf{r}, \mathbf{r}) \cdot \hat{m}(\mathbf{r})$ is called the shape anisotropy field, whose expression is actually more complicated due to the constraint $\hat{m}^2 = 1$; however we will prove that it is equivalent to the term here when used in LL equations. As discussed in Sect. 2.2, sometimes the micromagnetic is not a simple cube, but a polyhedron, then the demagnetizing matrix $\tilde{N}$ should be calculated following a formula similar to Eq. (2.43) by summing over the contributions of demagnetizing matrices of a rectangle surface and a triangle surface in Tables 2.2 and 2.3.

In Eq. (2.61), the magneto-elastic field due to interfacial stress is largely simplified. The H_{me} is the magneto-elastic field constant along an in-plane direction $\hat{e}_x$, and H_{me}' is related to the $m_x m_y$ energy term, if the thin film is in x–y plane.

2.3.2 LL Equation and LLG Equation

In 1935, Landau worked in the Haerkof University and the Physico-Technical Institute, Academy of Science, in Ukraine, USSR. Landau and his student Lifshitz brought up the famous LL equation in their theory of magnetic domain and domain wall resonance. The first term in LL equation can be derived from the Heisenberg equation of spin, which is related to the energy conservation of magnetic free energy. The second term in LL equation was said to be originated from the relativistic interaction between the magnetic moment and the crystal in the original paper [8]; nowadays this nonlinear second term is believed to be caused by the dissipative process, which is complicated and hard to be explained accurately, just similar to the frictional force (Fig. 2.6).

As discussed in Sect. 1.2, the atomic magnetic moment $\mu_a = -g\mu_B \mathbf{S}$, where g is the g-factor, μ_B is the Bohr magnet, $\mathbf{S}$ is the atomic spin. If we omit energy dissipation, the eigenvalue of Hamiltonian $\mathscr{H} = -\mu_a \cdot \mathbf{H}$, is conserved; then we can use the Heisenberg's equations to describe the motion of spin $\mathbf{S} = \hat{e}_i S_i$:

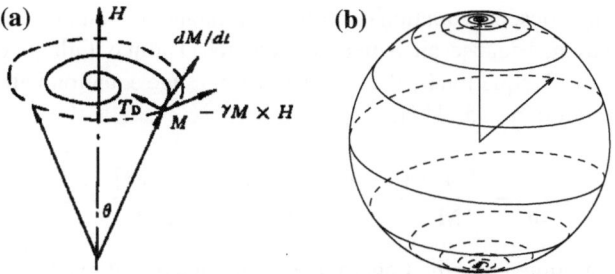

Fig. 2.6 Landau–Lifshitz equation. **a** Directions of precession term and damping term; **b** switching process of a magnetic moment or a spin from $+z$ to $-z$ direction by an external field

$$\frac{dS_i}{dt} = \frac{1}{i\hbar}[S_i,\,\mathcal{H}] = \frac{g\mu_B}{i\hbar}[S_i,\,S_j]\,H_j = \frac{g\mu_B}{\hbar}\varepsilon_{ijk}S_k H_j \tag{2.62}$$

where the commutator of the spin is $[S_i,\,S_j] = i\,\varepsilon_{ijk}S_k$. In a ferromagnetic material, near the 0 K, the atomic moment in a cell can be written as: $\mathbf{M} = n\mu = -ng\mu_B\mathbf{S}$, where n is the atomic density. Based on Eq. (2.62), if the energy is conserved, the equation of motion of magnetic moment is in the form:

$$\frac{d\mathbf{M}}{dt} = -ng\mu_B\frac{d\mathbf{S}}{dt} = -g\frac{e}{2mc}(\mathbf{M}\times\mathbf{H}) = -\gamma_0(\mathbf{M}\times\mathbf{H}) \tag{2.63}$$

where the constant $\gamma_0 = ge/2mc$ is the gyromagnetic ratio for an atomic spin. For the ferromagnetic alloys of Fe, Co, Ni transition metal elements, due to the quenching of orbital angular momentum of d-electrons, the g-factor of atomic spin usually is just the $g_0 = 2$ of an electron. Therefore the gyromagnetic ratio in the LL equation usually takes the value $\gamma_0 = e/mc = 1.75882 \times 10^7\,\mathrm{Oe^{-1}\,s^{-1}}$.

When the dissipation of magnetic free energy is included, the atomic spins have a non-equilibrium statistics, and its motion becomes nonlinear. Based on the Hamiltonian $\mathcal{H} = -\mu_a \cdot \mathbf{H}$, the dissipation of magnetic energy means the moment $\mathbf{M}$ will rotate to the direction of local magnetic field $\mathbf{H}$, which equals the effective magnetic field $\mathbf{H}_{\mathrm{eff}}$. Therefore Landau and Lifshitz added a damping term in the equation of motion of spin, which results in the famous LL equation:

$$\frac{d\mathbf{M}}{dt} = -\gamma(\mathbf{M}\times\mathbf{H}_{\mathrm{eff}}) - \gamma\frac{\alpha}{M}\mathbf{M}\times(\mathbf{M}\times\mathbf{H}_{\mathrm{eff}}) \tag{2.64}$$

where the dimensionless constant α is called the Landau damping constant, which reveals the dissipation speed in the ferromagnetic material. Similar to the frictional coefficient, the damping constant α is also phenomenological. Usually in Fe, Co, Ni metals or alloys, α is less than 0.1, sometimes even below 0.01. However, in ferromagnetic oxides or ferrites, the dissipation process is much slower, thus α can be one or two orders lower than the value in ferromagnetic metals.

In micromagnetics, the equation of motion for magnetic moments is always called the LLG equation, because an American scientist Thomas Gilbert explained the damping term of LL equation by the dissipative Lagrange equation with a Rayleigh's dissipation function in 1955 [12]:

$$\frac{\mathrm{d}}{\mathrm{d}t}\frac{\delta\mathscr{L}[\mathbf{M},\dot{\mathbf{M}}]}{\delta\dot{\mathbf{M}}} - \frac{\delta\mathscr{L}[\mathbf{M},\dot{\mathbf{M}}]}{\delta\mathbf{M}} + \frac{\delta\mathscr{R}[\dot{\mathbf{M}}]}{\delta\dot{\mathbf{M}}} = 0 \qquad (2.65)$$

where Gilbert assumed that the Lagrange equation itself will result in the Eq. (2.63) of motion for the magnetic moment under the constraint of energy conservation. In the Lagrange $\mathscr{L}[\mathbf{M},\dot{\mathbf{M}}] = \mathscr{T} - \mathscr{U}$, the role of magnetization $\mathbf{M}$ is the same as the position $\mathbf{r}$ in $\mathscr{L}[\mathbf{r},\dot{\mathbf{r}}]$ of classical mechanics. The Rayleigh dissipation functional is also constructed analogous to that of the frictional force in mechanics:

$$\mathscr{R}[\dot{\mathbf{r}}] = \frac{\eta}{2}\iiint \mathrm{d}^3\mathbf{r}\,\dot{\mathbf{r}}^2(\mathbf{r},t) \quad \Rightarrow \quad \mathscr{R}[\dot{\mathbf{M}}] = \frac{\eta}{2}\iiint \mathrm{d}^3\mathbf{r}\,\dot{\mathbf{M}}^2(\mathbf{r},t) \qquad (2.66)$$

Then the LLG equation can be obtained by inserting the Rayleigh dissipation functional into Eq. (2.65), where the effective field appears as $\delta\mathscr{U}/\delta\mathbf{M} = -\mathbf{H}_{\mathrm{eff}}$:

$$\frac{\mathrm{d}}{\mathrm{d}t}\frac{\delta\mathscr{T}[\mathbf{M},\dot{\mathbf{M}}]}{\delta\dot{\mathbf{M}}} - \frac{\delta\mathscr{T}[\mathbf{M},\dot{\mathbf{M}}]}{\delta\mathbf{M}} + \left(-\mathbf{H}_{\mathrm{eff}} + \eta\dot{\mathbf{M}}\right) = 0 \Rightarrow$$
$$\frac{\partial\mathbf{M}}{\partial t} = -\gamma_0\mathbf{M}\times\left(\mathbf{H}_{\mathrm{eff}} - \eta\dot{\mathbf{M}}\right) = -\gamma_0\mathbf{M}\times\mathbf{H}_{\mathrm{eff}} + \frac{\alpha}{M}\mathbf{M}\times\frac{\partial\mathbf{M}}{\partial t} \qquad (2.67)$$

where the damping $\alpha = \gamma_0\eta M$. The explicit expression for the kinetic energy $\mathscr{T}[\mathbf{M},\dot{\mathbf{M}}]$ of a rotating body is quite complicated, thus the derivation in Eq. (2.67) from the first line to the second line is obtained by the argument that this equation must be equivalent to Eq. (2.63) when the "friction coefficient" $\eta = 0$ [12].

The LLG equation and LL equation is totally equivalent to one another, except that the gyromagnetic ratio γ in the two equations has a small difference related to the damping. If we insert the right-hand side of the LLG equation in Eq. (2.67) into the last term $\partial\mathbf{M}/\partial t$, it is easy to prove that:

$$(1+\alpha^2)\frac{\partial\mathbf{M}}{\partial t} = -\gamma_0\mathbf{M}\times\mathbf{H}_{\mathrm{eff}} - \gamma_0\frac{\alpha}{M}\mathbf{M}\times(\mathbf{M}\times\mathbf{H}_{\mathrm{eff}}) \qquad (2.68)$$

Comparing this Eq. (2.68) with the LL Eq. (2.64), it is clear that:

$$\gamma = \gamma_0/(1+\alpha^2) \qquad (2.69)$$

Therefore the gyromagnetic ratio γ in the LL equation should be smaller than the gyromagnetic ratio γ_0 in the LLG equation, by a small factor related to the damping coefficient α.

2.3.3 History of Micromagnetics

The phrase "micromagnetics" was brought up by Brown in 1958, now it becomes the mainstream theory for computational applied magnetism. In Brown's book *Micromagnetics* [13], he summarized the "magnetization curve theory" and "domain theory" before 1960s, which are the main parts of today's micromagnetics. Furthermore, he studied the LL equation with the linear approximation or by the nonlinear calculation of static and dynamic problems.

The predecessor of micromagnetics was the magnetization-curve theory. The earliest theory of "induced magnetization" was given by W. E. Weber in 1852. The modern magnetization-curve theory was founded by Akulov and Becker. In 1929, Akulov derived the anisotropy energy for Fe and Ni with the cubic crystalline symmetry [9], as given in Eq. (2.54). To explain the magnetization curve of single crystals, as shown in Fig. 2.7a–c, a rigid rotation approximation can be used to explain the curve by minimizing the total energy $\mathscr{E}_a - \mathbf{H} \cdot \mathbf{M}$.

In Table 2.5, the expression of the anisotropy energy $\mathscr{E}_a$ and the derived magnetization curve, i.e. the $\bar{M} - H$ relationship, are listed in the hard axes for Fe, Co and Ni single crystal, respectively. It should be noted that the expressions of $\mathscr{E}_a$ for Fe and Ni are equivalent, except a constant, because both Fe and Ni have the cubic symmetry. M–H curves listed in Table 2.5 are also plotted in Fig. 2.7d–f.

If we compare the measured and the calculated magnetization curves under the rigid rotation approximation, several conclusions can be made. First, the remanence of M–H curves along hard axes can be estimated quite well. Secondly, the saturation field is on the order of the anisotropy field constant $H_k = 2K_1/M_s$; however, it can be seen in Fig. 2.7 and Table 2.5 that, in Fe, Co, Ni, the saturation field is about H_k, $0.15H_k$ and $0.65H_k$ respectively; therefore only the sample of Fe in Kaya's experiment [14] was near a single crystal. Most importantly, the M–H curve in the easy axis can not be explained by the rigid rotation model.

Thus Bloch's domain wall theory brought up during 1930–1931 was natural and necessary to explain these non-uniform rotation phenomena. A Bloch wall has a structure with zero pole density $\rho_M = -\nabla \cdot \mathbf{M}$; Therefore the magnetostatic interaction energy can be neglected in the free energy. Actually this was an approximation used in most of the early magnetization-curve theory and domain theory [8].

The M–H loop of single-domain particle was an important topic too in the magnetic theory. In 1930, Becker introduced the rigid rotational model, in conjunction with the concept of "internal stress", to explain the loops of an element in a magnetic material. The total free energy of this element was [10]:

$$U = U_{\text{dip}} - \mathbf{H} \cdot \mathbf{J} = -2SAJ^2 \cos(2(\theta - \varepsilon)) - HJ \cos\theta, \quad (2.70)$$

where S is a coefficient, A is the internal stress, J is the magnetization, θ is the angle between $\mathbf{J}$ and $\mathbf{H}$, and ε is the angle between $\mathbf{H}$ and the original position $\mathbf{B}$ of $\mathbf{J}$ when $\mathbf{H} = 0$ (along the stress anisotropy). Becker derived the M–H loop at different angle ε, where horizontal axis is H/H_σ with $H_\sigma = 4SAJ$, as seen in Fig. 2.8.

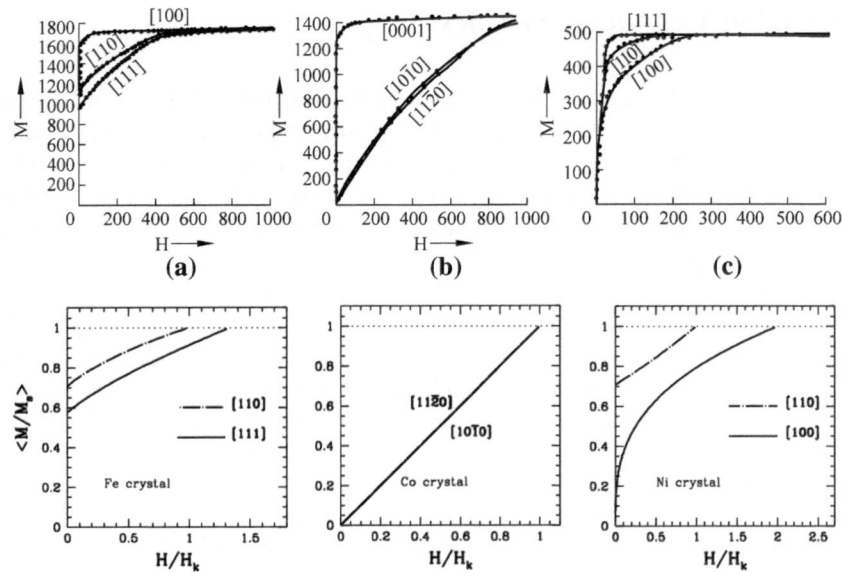

Fig. 2.7 a–c Measured M–H loops of Fe, Co, Ni [14]; **d–f** calculated M–H curves in the hard axes of single crystal Fe, Co, Ni under the rigid rotation approximation

Table 2.5 Magnetization curve $\bar{M} = \mathbf{M} \cdot \mathbf{H}/(M_s H)$ of Fe, Co, Ni by the rigid rotation model

Crystal	Fe	Co	Ni
$\mathscr{E}_a$	$2K_1(m_1^2 m_2^2 + m_2^2 m_3^2 + m_3^2 m_1^2)$	$K_1(m_1^2 + m_2^2)$	$-K_1(m_1^4 + m_2^4 + m_3^4)$
M_s	1742.6 emu/cm^3	1446.7 emu/cm^3	521.3 emu/cm^3
H_k	552 Oe	5696 Oe	−221 Oe
M–H 1	$H_{[110]}/H_k = \bar{M}(2\bar{M}^2 - 1)$	$H_{[11\bar{2}0]}/H_k = \bar{M}$	$H_{[110]}/H_k = (2\bar{M}^2 - 1)/\bar{M}$
M–H 2	$H_{[111]}/H_k = \dfrac{(3\bar{M}^3 - \bar{M})}{1 + (3\bar{M}^2 - 1)/4}$	$H_{[10\bar{1}0]}/H_k = \bar{M}$	$H_{[100]}/H_k = 2\bar{M}^3$

The famous Stoner–Wohlfarth model [15, 16] of a single-domain particle's M–H loop was mathematically equivalent to Becker's work. However, the physical explanation of the effective anisotropy energy of the single particle or element was different. In the first 50 years of magnetic recording, the magnetic media were granular or particulate media, which were composed of elongated particles such as γ-Fe$_2$O$_3$. The γ-Fe$_2$O$_3$ has a complicated cubic structure, and its crystalline anisotropy is small. In a recording medium, γ-Fe$_2$O$_3$ stores information due to its shape anisotropy, which was clarified by Stoner and Wohlfarth's work in 1948, and that is the reason why Stoner–Wohlfarth model is very important.

The effective anisotropy energy density of a Stoner–Wohlfarth single-domain particle is from the self-demagnetizing energy term as given in Eq. (2.51), where the self-demagnetizing matrix $\tilde{N}(\mathbf{r}, \mathbf{r})$ is diagonal, with $N_{11} = N_{22}$ and $N_{33} < N_{11}$ along its long-axis $\hat{k}_s$, due to the rotational symmetry of the elongated particle:

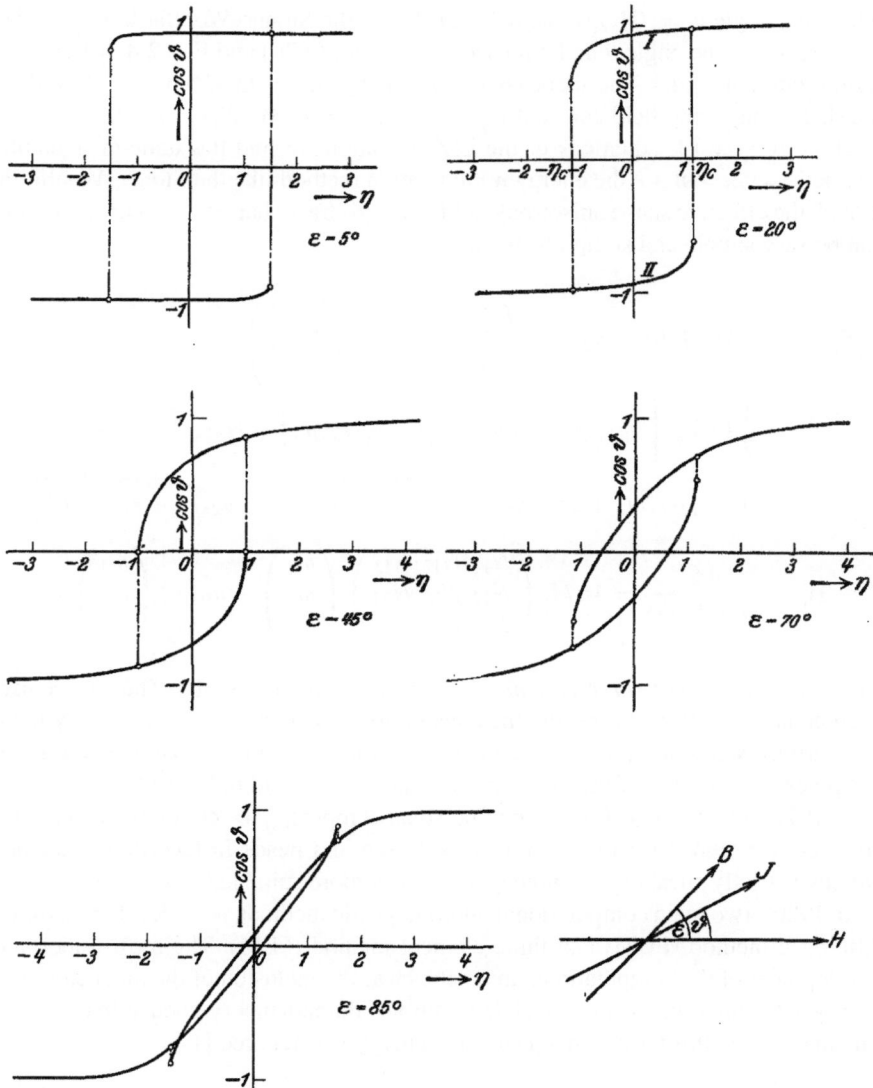

Fig. 2.8 M–H loops of an element with uniaxial anisotropy, where ε is the angle between the external field and the easy axis. © with kind permission from Springer Science and Business Media: [10], Fig. 3

$$\mathcal{E}/V = \frac{1}{2} 4\pi M_s^2 \left(N_{11}m_1^2 + N_{22}m_2^2 + N_{33}m_3^2 \right) - \mathbf{H} \cdot \mathbf{M}$$
$$= \mathcal{E}_0 - \left[2\pi M_s^2 (N_{11} - N_{33}) \right] (\hat{m} \cdot \hat{k}_s)^2 - \mathbf{H} \cdot \mathbf{M}$$
$$= \mathcal{E}_0 - K_s \cos^2(\phi - \theta) - M_s H \cos\phi \tag{2.71}$$

where the angle $\phi = \langle \mathbf{H}, \hat{m} \rangle$ and $\theta = \langle \mathbf{H}, \hat{k}_s \rangle$ of the Stoner–Wohlfarth model take the position of the angle θ and ε respectively in Eq. (2.70) and Fig. 2.8 of Becker's rigid rotational model. The shape anisotropy energy $K_s = 2\pi M_s^2 (N_{11} - N_{33})$; thus the shape anisotropy field constant $H_k^s = 2K_s/M_s = 4\pi M_s (N_{11} - N_{33})$.

If we use the LL equations or the LLG equations to find the static or dynamic magnetic states, but not the energy minimization method like the Stoner–Wohlfarth model, the effective shape anisotropy field of an arbitrary shaped micromagnetic cell can be very simple and straightforward:

$$\mathscr{E}_a^s/V = \frac{1}{2} 4\pi M_s^2 (m_1, m_2, m_3) \begin{pmatrix} N_{11} & N_{12} & N_{13} \\ N_{12} & N_{22} & N_{23} \\ N_{13} & N_{23} & N_{33} \end{pmatrix} \begin{pmatrix} m_1 \\ m_2 \\ m_3 \end{pmatrix} \qquad (2.72)$$

$$= \frac{1}{2} 4\pi M_s^2 \left\{ (N_{11} - N_{33})m_1^2 + (N_{22} - N_{33})m_2^2 + N_{33} \right.$$

$$\left. + 2N_{12}m_1m_2 + 2N_{13}m_1\sqrt{1 - m_1^2 - m_2^2} + 2N_{23}m_2\sqrt{1 - m_1^2 - m_2^2} \right\}$$

$$-\mathbf{H}_k^s = \frac{\partial \mathscr{E}_a^s}{\partial (V M_s \hat{m})} = 4\pi M_s \begin{pmatrix} N_{11} & N_{12} & N_{13} \\ N_{12} & N_{22} & N_{23} \\ N_{13} & N_{23} & N_{33} \end{pmatrix} \begin{pmatrix} m_1 \\ m_2 \\ m_3 \end{pmatrix} - h_0\{\hat{m}\} \begin{pmatrix} m_1 \\ m_2 \\ m_3 \end{pmatrix} \quad (2.73)$$

where $h_0\{\hat{m}\} = 4\pi M_s (m_1 N_{13} + m_2 N_{23} + m_3 N_{33})/m_3$ is a scalar. Thus, if we use LLG equations, after considering the constraint $\hat{m}^2 = 1$, the shape anisotropy field of an arbitrary cell in Eq. (2.73) is equivalent to the self-demagnetizing field term in the general expression of the magnetostatic interaction field in Eq. (2.60).

In 1970s, owing to the invention of personal computers, the computational methods were developed gradually to analyze media and heads in recording systems, although strictly speaking, "nanomagnetics" is a more suitable term.

In 1980s, two main computational micromagnetic methods were developed: finite difference method (FDM) and finite element method (FEM). The most important development of the computational magnetics was the inclusion of the magnetostatic energy in the micromagnetic model. In traditional domain theory, such as in Landau–Lifshitz's work, this term of magnetostatic energy was ignored [1]:

$$\mathscr{E} = \int dV \left[\frac{1}{2}\alpha s'^2 + \frac{1}{2}\beta \left(s_x^2 + s_y^2 \right) \right], \quad \mathbf{s} = (0, s\sin\theta, s\cos\theta), \quad (2.74)$$

$$\alpha\theta'' - \beta\sin\theta\cos\theta = 0 \quad \Rightarrow \quad \cos\theta = -\tanh[x/\sqrt{\alpha/\beta}] \quad (2.75)$$

where the coefficient $\alpha = 2A^*$, $\beta = 2K_1$, and $s'^2 = \sum_i (ds_i/dx)(ds_i/dx)$. The boundary condition of the Bloch wall is $\theta = 0, \pi$ at $x = -\infty, +\infty$, respectively. The Bloch wall width $a = \sqrt{\alpha/\beta}$ equals the Bloch exchange length $l_{ex}^B = \sqrt{A^*/K_1}$. The ignore of magnetostatic energy was reasonable for the Bloch wall, because its magnetic pole density $-\nabla \cdot \mathbf{M}$ is zero. However, for most of other cases in

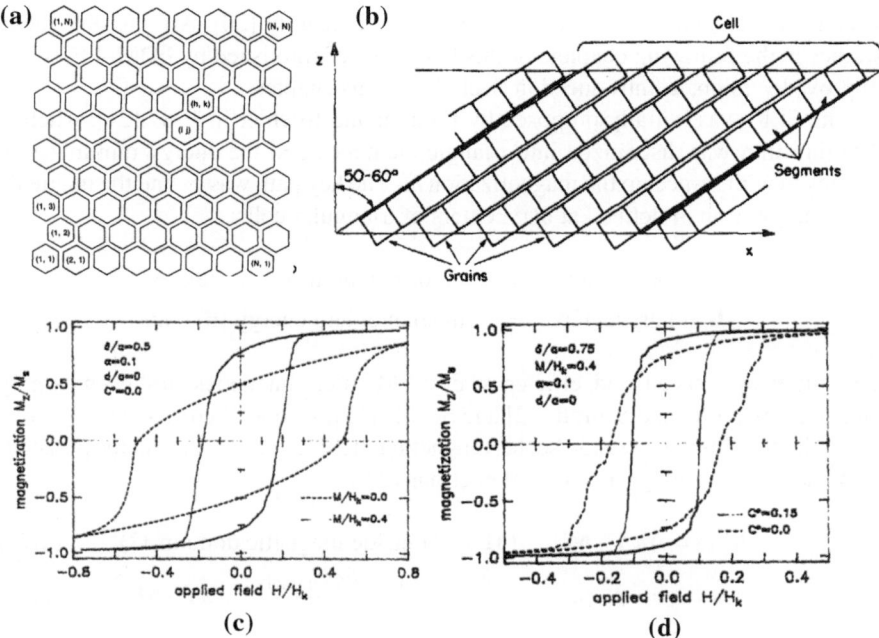

Fig. 2.9 FDM and FDM–FFT micromagnetic method. **a** Model of magnetic grains in Co–P thin film; © Reprinted with permission from Ref. [17]. Copyright [1983], American Institute of Physics; **b** model of micromagnetic cells in CoNi film; © Reprinted with permission from Ref. [18]. Copyright [1987], American Institute of Physics; **c, d** effects of magnetostatic interaction (M_s/H_k) and exchange interaction ($C^* = H_e/H_k$) among the grains on M–H loops using FDM–FFT method; © Reprinted with permission from Ref. [19]. Copyright [1988], American Institute of Physics

applied magnetism, this ignorance of the demagnetizing field is not correct, and the computational magnetics is necessary to include the term $\mathbf{H}_m$ in Eq. (2.60).

The FDM uses regular mesh of micromagnetic cells to discretize the magnetic material. In 1983, Hughes used the energy minimization method to calculate the M–H loops of Co–P thin film, where hexagonal grains arranged on a triangular lattice was chosen as the regular mesh, as seen in Fig. 2.9a [17]. In 1987, Victora used the LLG equations to calculate the torque curve of CoNi film, where the complete magnetostatic interaction was included for micromagnetic cells in Fig. 2.9b by a direct integration [18]. In 1988, Bertram and Zhu developed a FDM–FFT method for the micromagnetic model, where the computation of the magnetostatic interaction was largely speed up by using the fast-Fourier-transform (FFT) method [19]. By using FDM–FFT method, the effects of the magnetostatic interaction among all grains and the exchange among neighbor grains can be carefully studied, as seen in Fig. 2.9c, d, which was important for the development of low-noise thin film recording media. The Object-Oriented MicroMagnetic Framework (OOMMF) software of micromagnetics was initiated in the Intermag 1995, based on discussions of micromagnetic standard problems in Intermag 1994 and Intermag 1995. OOMMF uses FDM–FFT methods, which is developed by NIST and widely utilized recently [20]. All micromagnetic studies in the following chapters of this book are accomplished by FDM–FFT or its improved version, as introduced in Sect. 2.2 of this chapter.

The FEM was initially proposed by Fredkin and Koehler in 1987 [21], where a 2-D thin film was discretized into triangle elements, and the energy minimization was taken with respect to the magnetization $\hat{m}$. The key path was to find the magnetic field energy from $\hat{m}$ defined at barycenters of triangular cells:

$$\hat{m} \text{ at barycenter} \quad \Rightarrow \quad \text{vector potential } \mathbf{A} \text{ at vertex} \Rightarrow$$
$$\mathbf{B} \text{ at barycenter} \quad \Rightarrow \quad \text{magnetic field energy } |\mathbf{B}|^2/8\pi \qquad (2.76)$$

The simulation speed and efficiency of FEM micromagnetics has been largely improved by Fidler and Schrefl [22] and Scholz [23], who worked at Vienna University of Technology, where the scalar magnetic potential but not the vector potential is the key to solve magnetostatic interactions [23]:

$$\nabla^2 \phi = 4\pi \nabla \cdot \mathbf{M} \qquad (\mathbf{M} = 0 \text{ outside magnetic domain } \Omega) \qquad (2.77)$$

$$\phi_{\text{external}} = \phi_{\text{internal}} , \qquad \frac{\partial \phi_{\text{external}}}{\partial \hat{n}} - \frac{\partial \phi_{\text{internal}}}{\partial \hat{n}} = 4\pi \hat{n} \cdot \mathbf{M} \qquad (2.78)$$

The Dirichilet boundary condition in Eq. (2.78) contains an approximation, which will be better in soft magnetic materials than in hard magnetic materials.

Appendix

Appendix A

If a $a' \times b'$ rectangle locates at $z' = 0$, and the observation position vector is located at $\mathbf{r} = (x, y, z) = (r_1, r_2, r_3)$, as seen in Fig. 2.4a, the three nonzero elements in the demagnetizing matrix of a rectangular surface are ($\alpha = 1, 3$):

$$N_{\alpha 3}^{\text{rec}}(\mathbf{r}) = -\frac{1}{4\pi} \int\limits_{-a'/2}^{a'/2} dr_1' \int\limits_{-b'/2}^{b'/2} dr_2' \frac{r_\alpha - r_\alpha'}{\left[(r_1 - r_1')^2 + (r_2 - r_2')^2 + r_3^2\right]^{3/2}} \qquad (2.79)$$

We just have to do two integrations for N_{33}^{rec} and N_{13}^{rec}, and the integration of N_{23}^{rec} is totally analogy to N_{13}^{rec}. Let's start with N_{13}^{rec}:

$$N_{13}^{\text{rec}} = -\frac{1}{4\pi} \int\limits_{-a'/2}^{a'/2} dx' \int\limits_{-b'/2}^{b'/2} dy' \frac{x - x'}{\left[(x - x')^2 + (y - y')^2 + z^2\right]^{3/2}}$$

$$= -\frac{1}{4\pi} \int\limits_{-b'/2}^{b'/2} dy' \frac{1}{\left[(x - x')^2 + (y - y')^2 + z^2\right]^{1/2}} \Bigg|_{x'=-a'/2}^{a'/2}$$

$$= -\frac{1}{4\pi} \sum_q (-q) \ln\left(y' - y + \sqrt{R_1^2 + (y - y')^2 + R_3^2}\right)\Bigg|_{y'=-b'/2}^{b'/2}$$

$$= -\frac{1}{4\pi} \sum_q \sum_w qw \ln\left(R - wR_2\right) \qquad (2.80)$$

The variables are defined in Table 2.2. The integration for N_{33}^{rec} can also be done:

$$N_{33}^{\text{rec}} = -\frac{1}{4\pi} \int\limits_{-a'/2}^{a'/2} dx' \int\limits_{-b'/2}^{b'/2} dy' \frac{z}{\left[(x - x')^2 + (y - y')^2 + z^2\right]^{3/2}}$$

$$= -\frac{1}{4\pi} \int\limits_{-b'/2}^{b'/2} dy' \frac{z}{\left[(y - y')^2 + z^2\right]} \frac{x' - x}{\left[(x - x')^2 + (y - y')^2 + z^2\right]^{1/2}} \Bigg|_{x'=-a'/2}^{a'/2}$$

$$= -\frac{1}{4\pi} \sum_q \arctan\left(\frac{R_1}{R_3} \frac{y' - y}{\sqrt{R_1^2 + (y - y')^2 + R_3^2}}\right)\Bigg|_{y'=-b'/2}^{b'/2}$$

$$= -\frac{1}{4\pi} \sum_q \sum_w \arctan \frac{R_1 R_2}{R_3 R} \qquad (2.81)$$

The $N_{\alpha 3}^{\text{rec}}(\mathbf{r})$ of a rectangular surface have be listed in Table 2.2 respectively.

Appendix B

If the surface located at $z' = 0$ is a right-angle triangle with right-angle side lengths (a', b'), the origin at the midpoint of the hypotenuse, and the hypotenuse defined by equation $y' = -sx'$ or $x' = -vy'$, and the observation position vector is located at $\mathbf{r} = (x, y, z)$, there are still three nonzero elements N_{13}^{tri}, N_{23}^{tri} and N_{33}^{tri} in the demagnetizing matrix of a triangular surface, as seen in Fig. 2.4b. Here we can first do the integral for element N_{13}^{tri}:

$$N_{13}^{\mathrm{tri}} = -\frac{1}{4\pi} \int\limits_{-b'/2}^{b'/2} dy' \int\limits_{-vy'}^{a'/2} dx' \frac{x - x'}{\left[(x - x')^2 + (y - y')^2 + z^2\right]^{3/2}}$$

$$= -\frac{1}{4\pi} \int\limits_{-b'/2}^{b'/2} dy' \left\{ \frac{1}{\left[(a'/2 - x)^2 + (y - y')^2 + z^2\right]^{1/2}} \right.$$

$$\left. - \frac{1}{\left[(x + vy')^2 + (y - y')^2 + z^2\right]^{1/2}} \right\}$$

$$= -\frac{1}{4\pi} \ln \left(y' - y + \sqrt{(a'/2 - x)^2 + (y - y')^2 + z^2} \right) \Bigg|_{y'=-b'/2}^{b'/2}$$

$$+ \frac{1}{4\pi} \int\limits_{-b'/2}^{b'/2} dy' \frac{1}{\sqrt{1 + v^2}} \frac{1}{\left[(y' - c_1)^2 + r^2/(1 + v^2) - c_1^2\right]^{1/2}}$$

$$= \frac{1}{4\pi} \sum_w w \ln (R_{\mathrm{I}} - w R_2)$$

$$+ \frac{1}{4\pi} \frac{1}{\sqrt{1 + v^2}} \ln \left(y' - c_1 + \sqrt{(y' - c_1)^2 + c_2^2} \right) \Bigg|_{y'=-b'/2}^{b'/2}$$

$$= -\frac{1}{4\pi} \sum_w w \left\{ \frac{1}{\sqrt{1 + v^2}} \ln(P - w P_2) - \ln(R_{\mathrm{I}} - w R_2) \right\} \tag{2.82}$$

The symbols c_1, c_2^2, P, P_2, R_{I}, and R_2 used in Eq. (2.82) have be defined in Tables 2.2 and 2.3 respectively. The integration for N_{23}^{tri} is totally analogy to N_{13}^{tri}, just with a $x \leftrightarrow y$ symmetry, therefore the derivation of N_{23}^{tri} will be omitted here.

The integral for the matrix element N_{33}^{tri} is the most complicated one, which include three parts:

$$N_{33}^{\mathrm{tri}} = -\frac{1}{4\pi} \int\limits_{-b'/2}^{b'/2} dy' \int\limits_{-vy'}^{a'/2} dx' \frac{z}{\left[(x - x')^2 + (y - y')^2 + z^2\right]^{3/2}}$$

$$= -\frac{1}{4\pi} \int\limits_{-b'/2}^{b'/2} dy' \frac{z}{\left[(y - y')^2 + z^2\right]} \frac{x' - x}{\left[(x - x')^2 + (y - y')^2 + z^2\right]^{1/2}} \Bigg|_{x'=-vy'}^{a'/2}$$

$$= -\frac{1}{4\pi} \left\{ N_{33}^{(1)} + N_{33}^{(2)} + N_{33}^{(3)} \right\} \tag{2.83}$$

In Eq. (2.83), the derivation of the first term $N_{33}^{(1)}$ is actually very similar to one of the two terms in Eq. (2.81) for rectangular surface; in the second term, the numerator of the integrand is $z(-vy' - x)$, which can be disassembled into two parts $zv(-y' + y)$ and $z(-vy - x)$, and these two part just corresponds to the $N_{33}^{(2)}$ and $N_{33}^{(3)}$ respectively.

The integration for $N_{33}^{(1)}$ is just straightforward:

$$N_{33}^{(1)} = \int_{-b'/2}^{b'/2} dy' \frac{z}{[(y-y')^2 + z^2]} \frac{a'/2 - x}{[(a'/2 - x)^2 + (y - y')^2 + z^2]^{1/2}}$$

$$= \arctan\left(\frac{a'/2 - x}{z} \frac{y' - y}{\sqrt{(a'/2 - x)^2 + (y - y')^2 + z^2}}\right)\Bigg|_{y'=-b'/2}^{|b'/2}$$

$$= \sum_w \arctan[(a'/2 - x)R_2/(zR_1)] \tag{2.84}$$

In the derivation for $N_{33}^{(2)}$, complicated variables such as c_1, c_2 and $c_5 = y - iz$ in Table 2.3 have to be defined, and the respective integral is:

$$N_{33}^{(2)} = \int_{-b'/2}^{b'/2} dy' \frac{zv}{[(y-y')^2 + z^2]} \frac{y' - y}{[(x + vy')^2 + (y - y')^2 + z^2]^{1/2}} \tag{2.85}$$

$$= \int_{-b'/2}^{b'/2} dy' \frac{zv}{2} \left[\frac{1}{y' - c_5} + \frac{1}{y' - c_5^*}\right] \frac{1}{\sqrt{1 + v^2}} \frac{1}{[(y' - c_1)^2 + c_2^2]^{1/2}}$$

The previous integration include two terms which are complex conjugates of one another. Now let's make an integration variable change $y' - c_1 = c_2 \sinh\theta$ with the two integration limits of the angle θ as $\theta_1 = \sinh^{-1}[(-b'/2 - c_1)/c_2]$ and $\theta_2 = \sinh^{-1}[(b'/2 - c_1)/c_2]$, the integral in Eq. (2.85) has the form:

$$N_{33}^{(2)} = \frac{zv}{2\sqrt{1 + v^2}} \int_{\theta_1}^{\theta_2} \left[\frac{d\theta}{c_1 - c_5 + c_2 \sinh\theta} + c.c.\right] \tag{2.86}$$

Then make another variable change $e^\theta = u$, and define a new constant $\sinh\eta = (c_1 - c_5)/c_2$, the integral becomes:

$$N_{33}^{(2)} = \frac{zv}{\sqrt{1+v^2}} \int_{u_1}^{u_2} \left[\frac{1}{c_2} \frac{d\left(e^\theta\right)}{\left(e^\theta\right)^2 + 2[(c_1 - c_5)/c_2]e^\theta - 1} + c.c. \right]$$

$$= \frac{zv}{\sqrt{1+v^2}} \int_{u_1}^{u_2} \left[\frac{1}{c_2} \frac{du}{(u + e^\eta)(u - e^{-\eta})} + c.c. \right]$$

$$= \frac{zv}{\sqrt{1+v^2}} \left\{ \frac{1}{2c_2 \cosh \eta} \ln \frac{u - e^{-\eta}}{u + e^\eta} \Big|_{u=u_1}^{u_2} + c.c. \right\}$$

$$= \frac{zv}{\sqrt{1+v^2}} \sum_w \Re \left\{ \frac{w}{\sqrt{(c_1 - c_5)^2 + c_2^2}} \ln \frac{V_+}{V_-} \right\}$$

$$= \frac{zv}{\sqrt{1+v^2}} \sum_w \Re \left\{ \frac{w}{A e^{i\theta'/2}} \ln \frac{|V_+| e^{i\phi_+}}{|V_-| e^{i\phi_-}} \right\} \qquad (2.87)$$

In the previous derivation, the difficult part is to find integration limits u_1 and u_2. Actually $\sinh^{-1} x = \ln[x + \sqrt{x^2 + 1}]$, therefore u_1, u_2, e^η and $e^{-\eta}$ are:

$$u_1 = e^{\theta_1} = \frac{1}{c_2} \left[-\left(\frac{b'}{2} + c_1\right) + \sqrt{\left(\frac{b'}{2} + c_1\right)^2 + c_2^2} \right]$$

$$u_2 = e^{\theta_2} = \frac{1}{c_2} \left[+\left(\frac{b'}{2} - c_1\right) + \sqrt{\left(\frac{b'}{2} - c_1\right)^2 + c_2^2} \right]$$

$$e^\eta = \frac{1}{c_2} \left[+(c_1 - c_5) + \sqrt{(c_1 - c_5)^2 + c_2^2} \right]$$

$$e^{-\eta} = \frac{1}{c_2} \left[-(c_1 - c_5) + \sqrt{(c_1 - c_5)^2 + c_2^2} \right] \qquad (2.88)$$

By defining $P_2 = \frac{b'}{2} + wc_1$ ($w = +1$ and $w = -1$ are for the two integral limits u_1 and u_2 respectively), $A e^{i\theta'/2}$, V_+ and V_- will have the forms in Table 2.3:

$$\frac{u - e^{-\eta}}{u + e^\eta} = \frac{-wP_2 + \sqrt{P_2^2 + c_2^2} + (c_1 - c_5) - \sqrt{(c_1 - c_5)^2 + c_2^2}}{-wP_2 + \sqrt{P_2^2 + c_2^2} + (c_1 - c_5) + \sqrt{(c_1 - c_5)^2 + c_2^2}} = \frac{V_-}{V_+} (2.89)$$

The integration of $N_{33}^{(3)}$ is similar to $N_{33}^{(2)}$, which also includes imaginary numbers:

$$N_{33}^{(3)} = \int_{-b'/2}^{b'/2} dy' \frac{z}{\left[(y-y')^2+z^2\right]} \frac{x+vy}{\left[(x+vy')^2+(y-y')^2+z^2\right]^{1/2}} \qquad (2.90)$$

$$= \int_{-b'/2}^{b'/2} dy' \frac{x+vy}{2i} \left[\frac{-1}{y'-c_5} + \frac{1}{y'-c_5^*}\right] \frac{1}{\sqrt{1+v^2}} \frac{1}{\left[(y'-c_1)^2+c_2^2\right]^{1/2}}$$

The rest of derivations are similar to Eq. (2.87), but the result is an imaginary part:

$$N_{33}^{(3)} = -\frac{x+vy}{2i\sqrt{1+v^2}} \int_{\theta_1}^{\theta_2} \left[\frac{d\theta}{c_1-c_5+c_2\sinh\theta} - c.c.\right]$$

$$= -\frac{x+vy}{2i\sqrt{1+v^2}} \int_{u_1}^{u_2} \left[\frac{1}{c_2}\frac{du}{(u+e^\eta)(u-e^{-\eta})} - c.c.\right]$$

$$= -\frac{x+vy}{\sqrt{1+v^2}} \sum_w \Im \left\{\frac{w}{\sqrt{(c_1-c_5)^2+c_2^2}} \ln \frac{V_+}{V_-}\right\}$$

$$= -\frac{x+vy}{\sqrt{1+v^2}} \sum_w \Im \left\{\frac{w}{Ae^{i\theta'/2}} \ln \frac{|V_+|e^{i\phi_+}}{|V_-|e^{i\phi_-}}\right\} \qquad (2.91)$$

Finally, insert the results of the three parts in Eqs. (2.84), (2.87) and (2.91) into Eq. (2.81), the most difficult matrix element of a triangular surface N_{33}^{tri} can be found.

References

1. Landau, L.D., Lifshitz, E.: Electrodynamics of Continuous Media, translated from Russian by Sykes J.B. and Bell J.S. Pergamon Press, Oxford (1960)
2. Brown, W.F., Jr., La Bonte, A.E.: Structure and energy of one-dimensional domain walls in ferromagnetic thin films. J. Appl. Phys. **36**(4), 1380–1386 (1965)
3. Maxwell J.C.: A Treatise on Electricity and Magnetism (1873), translated by Ge G. into Chinese. Wuhan Press, Wuhan (1994)
4. von Laue M.: Geschichte der Physik (1950), translated by Fan D. N. and Dai N.Z. into Chinese. Commercial Press, Beijing (1978)
5. Wei, D., Wang, S.M., Ding, Z.J., Gao, K.Z.: Micromagnetics of ferromagnetic nano-devices using fast fourier transform method. IEEE Trans. Magn. **45**(8), 3035–3045 (2009)
6. Schabes, M.E., Aharoni, A.: Magnetostatic interaction fields for a three-dimensional array of ferromagnetic cubes. IEEE Trans. Magn. **23**(6), 3882–3888 (1987)
7. Wei, D.: Fundamentals of Electric, Magnetic, Optic Materials and Devices (in chinese), 2nd edn. Science Press, Beijing (2009)

8. Landau, L.D., Lifshitz, E.: On the theory of the dispersion of magnetic permeability in ferro-magnetic bodies. Phys. Zeitsch. der Sow. **8**, 153 (1935), reprinted in English by Ukr. J. Phys. **53**, 14 (2008)
9. Akulov, N.S.: Zur atomtheorie des ferromagnetismus. Z. Phys. **54**, 582–587 (1929)
10. Becker, R.: Zur theorie der magnetisierungskurve. Z. Phys. **62**, 253–269 (1930)
11. Bloch, F.: Zur theorie des Austauschproblems und der Remanenzerscheinung der ferromag-netika. Z. Phys. **74**, 295–335 (1932)
12. Gilbert, T.L.: A phenomenological theory of damping in ferromagnetic materials. IEEE Trans. Magn. **40**(6), 3443–3449 (2004)
13. Brown, W.F., Jr.: Micromagnetics. Wiley, New York (1963)
14. Kaya, S.: On the magnetization of single crystals of nickel, vol. 17, p. 1157. Science Report, Tohoku University (1928)
15. Stoner, E.C., Wohlfarth, E.P.: A mechanism of magnetic hysteresis in heterogeneous alloys. IEEE Trans. Magn. **27**(4), 3475–3518 (1991)
16. Stoner, E.C., Wohlfarth, E.P.: A mechanism of magnetic hysteresis in heterogeneous alloys. Philos. Trans. R. Soc. Lond. A **240**, 599 (1948)
17. Hughes, G.F.: Magnetization reversal in cobalt-phosphorus films. J. Appl. Phys. **54**, 5306–5313 (1983)
18. Victora, R.H.: Micromagnetic predictions for magnetization reversal in CoNi films. J. Appl. Phys. **62**, 4220–4225 (1987)
19. Bertram, H.N., Zhu, J.G.: Micromagnetic studies of thin metallic films. J. Appl. Phys. **63**, 3248–3253 (1988)
20. NIST: Object Oriented MicroMagnetic Framework (OOMMF) Project. NIST Center for Infor-mation Technology Laboratory. http://www.math.nist.gov/oommf/ (2006). Accessed 20 June 2011
21. Fredkin, D.R., Koehler, T.R.: Numerical micromagnetics by the finite element method. IEEE Trans. Magn. **23**(5), 3385–3387 (1987)
22. Fidler, J., Schrefl, T.: Micromagnetic modelling—the current state of the art. J. Phys. D Appl. Phys. **33**, R135–R156 (2000)
23. Scholz, W., Fidler, J., Schrefl, T., Suess, D., Dittrich, R., Forster, H., Tsiantos, V.: Scalable parallel micromagnetic solvers for magnetic nanostructures. Comput. Mater. Sci. **28**, 366–383 (2003)

Chapter 3
Microstructure and Hysteresis Loop

Abstract This chapter will discuss the micromagnetic simulation of hysteresis loops or M–H loops of magnetic thin films, based on their microstructures. The practical thin films in magnetic recording industry are all polycrystalline; therefore the microstructure of a thin film has multi-scales, such as defects, crystal grains, grain boundary, atomic symmetries in the grain and at the grain boundary. The Landau–Lifshitz equations are utilized to calculate the M–H loops, since this is a quasi-static process, the damping constant can be relatively large. The structure of programs for micromagnetic simulation will also be discussed in this chapter.

Keywords Programming in micromagnetics · Anisotropy distribution · Microstructure simulation · Perpendicular recording media · Soft magnetic thin film · Tunneling magnetoresistive spin valve

The manufacturing of CoCr thin film media in hard disk drives was accomplished by Robert I. Potter and Niel Heiman in LANX in early 1980s. The simulations of M–H loops in Co-alloy thin film hard magnetic media were developed thereafter. In 1983, Hughes in Seagate Technology used the energy minimization method to calculate the M–H loops of Co–P thin film, using a model with hexagonal grains arranged on a triangular lattice, as seen in Fig. 2.9a. In Hughes' work, although approximations were used, the effect of the magnetostatic interactions among grains on M–H loops were clarified: the squareness increases and the coercivity decreases [1]. In 1987, Victora used the LLG equations to calculate the torque curve of CoNi film, where the complete magnetostatic interaction was included for micromagnetic cells in Fig. 2.9b by a direct integration. Thus the simulation of the torque curve fitted very well with the experiment [2].

In 1988, Bertram and Zhu developed a computational method for the micromagnetic model, where the calculation of the magnetostatic interaction was largely sped up by using the fast-Fourier-transform (FFT) method [3]. By using this

---

An erratum to this chapter is available at 10.1007/978-3-642-28577-6_5.

FDM–FFT method, the effects of the magnetostatic interaction among all grains and the exchange among neighbor grains can be carefully studied. Their work pointed out the key to develop a low-noise longitudinal thin film recording media: the magnetostatic interaction and the exchange interaction should both be low among grains. In the longitudinal CoCrX thin film media in 1980s, the segregation of the nonmagnetic Cr-rich phase on the grain boundary was naturally formed in the sputtering process: the M_s was relatively low to decrease the demagnetizing field and the inter-granular exchange A^* was also low due to the Cr-rich phase.

In this chapter, micromagnetic models were built up based on the microstructures of polycrystalline magnetic thin films such as thin film media, FeCo thin films and tunneling magnetoresistive (TMR) multilayers utilized in storage, write and read magnetic devices of hard disk drives. A regular mesh, cubic or tetragonal, is chosen for micromagnetic cells; therefore the FFT and periodic boundary condition can be performed easily and correctly. The crystal grains are simulated on this regular mesh by grain-growth simulation around the initial nuclei; and grain boundaries are set between neighbor grains. The atomic lattice symmetry in the phase of crystal grains would be considered for different magnetic materials. The Landau–Lifshitz equations are utilized to calculate the M–H loops, since this is a quasi-static process, the damping constant can be relatively large. The structure of programs for micromagnetic simulation will also be discussed in this chapter.

3.1 Program Structure for Micromagnetics

The hysteresis loop or M–H loop comprehensively reflects the macroscopic properties of ferromagnetic materials. Before the studies of devices and read/write process in hard disk drives, the M–H loops should be analyzed to obtain the appropriate model of the thin films. The procedure of simulation is given in Table 3.1.

The principle of science is "The test of all knowledge is experiment" [4]. So is the micromagnetics. In the micromagnetic simulation of M–H loops, we need to input the geometric and intrinsic magnetic parameters, some directly from the experiments, some just by guesses. Finally if the simulation can explain the experiment well, we will know that the choices of the "guessed" parameters are correct, and these parameters can be used in advanced models.

The geometric parameters describe the shape and microstructure of the thin film, such as: film thickness δ, grain size D_g, and scales of other larger defects. The micromagnetic cell size a_L or D of the regular mesh is also an input geometric parameter, however the choice of D is a complicated problem, and we will briefly discuss here and further discuss this topic in next chapter.

There are two intrinsic scales in a magnetic material: Bloch exchange length $l_{ex}^B = \sqrt{A^*/K_1}$, the Bloch wall width; and Néel exchange length $l_{ex}^N = \sqrt{A^*/2\pi M_s^2}$, the Néel wall width. In a soft magnetic material such as FeCo, l_{ex}^B is large ($\sim 100\,\text{nm}$) and l_{ex}^N is very small ($\sim \text{nm}$), usually we can chose $D \sim l_{ex}^N$ or even larger than l_{ex}^N; in

Table 3.1 Procedures of micromagnetic simulations

| Assumptions | (1) In a micromagnetic cell the magnetic moment rotates uniformly |
	(2) Intrinsic parameters M_s, H_k, A^* of all micromagnetic cells are inputs
Step 1	Chose appropriate geometric model based on microstructure of thin film
Step 2	Chose appropriate free energy terms for micromagnetic cells in the model
Step 3	Calculate effective magnetic field terms based on free energy expression
Step 4	Solve Landau–Lifshitz equations for all micromagnetic cells numerically
Step 5	Based on the request, design the external field $\mathbf{H}_{ext}(\mathbf{r}, t)$, then output

Table 3.2 Structure of micromagnetic programs

Main.f	Main program: input, change $\mathbf{H}_{ext}$, find $\mathbf{H}_{eff}$, solve LL equations, output
Grains.f	Define microstructure of thin film, growing the grains and set grain boundary
Anisodis.f	Set $\hat{k}$, H_k based on orientation and magnitude distribution $f(\theta)$, $P(H_k)$
DemagM.f	Find demagnetizing matrix $\tilde{N}(\mathbf{r}, \mathbf{r}_0)$ between all cells $\mathbf{r}$ and a fixed cell $\mathbf{r}_0$
fftDemag.f	Calculate the FFT of demagnetizing matrix $\tilde{N}(\mathbf{k})$ at any wave vector $\mathbf{k}$
AnisoH.f	In each step of solving LL equations, find anisotropy fields $\mathbf{H}_a$ for all cells
ExchaH.f	In each step of solving LL equations, find exchange fields $\mathbf{H}_{ex}$ for all cells
DemagH.f	In each step of solving LL equations, find demagnetizing fields $\mathbf{H}_m$ for all cells

a hard magnetic medium such as Co-alloy, l_{ex}^B is small (<10 nm) and l_{ex}^N is also not large (~ 10 nm), the cell size D is usually smaller than both l_{ex}^B and l_{ex}^N.

The intrinsic magnetic parameters describe the spontaneous magnetization of a thin film, such as: magnetization M_s, anisotropy field $H_k = 2K_1/M_s$, exchange constant A^*. Due to the complexity of microstructure in polycrystalline thin film, we usually will use two sets of magnetic parameters: one for cells in the crystalline phase inside a magnetic grain, one for cells in the amorphous phase at the grain boundary. In a hard magnetic recording medium, the phase at grain boundary is nonmagnetic, where $M_s = 0$ can be set for all boundary cells; however, we still need two exchange constant: A_1^* for exchange between two cells in a grain, A_2^* for exchange across the grain boundary; in a soft magnetic thin film or in the GMR/TMR multilayers, the amorphous phase at grain boundary is magnetic, and we should chose M_s, H_k, A_1^*, and A_2^* carefully based on the physics in experiments.

An example of the structure of a micromagnetic program is given in Table 3.2. The main program "main.f" can be viewed as a controller and a solver. The four subroutines "Grains.f", "Anisodis.f", "DemagM.f" and "fftDemag.f" only have to be called once, which prepare the microstructure of the simulated media and the Fourier transform of the demagnetizing matrix in a regular mesh, respectively. The three subroutines "AnisoH.f", "ExchaH.f" and "DemagH.f" calculate the effective field terms, which will be called thousands of times when solving the Landau–Lifshitz equations; therefore they have to be written neatly to minimize the running time. The specific programs are omitted due to the length limit of the book.

3.1.1 FFT and Periodic Boundary Condition

In the calculation of the demagnetizing field, an infinite or periodic regular mesh has to be employed, because $\tilde{N}(\mathbf{r}_i, \mathbf{r}_j) = \tilde{N}(\mathbf{r}_i - \mathbf{r}_j, 0)$ is required for FFT. Following Eq. (2.60), the demagnetizing field in an arbitrary cell i can be calculated:

$$
\begin{aligned}
\mathbf{H}_{\mathrm{m}}(\mathbf{r}_i) &= -(4\pi M_{\mathrm{s}}) \sum_j \tilde{N}(\mathbf{r}_i, \mathbf{r}_j) \cdot \hat{m}_j \\
&= -(4\pi M_{\mathrm{s}}) \sum_j \frac{1}{N_{\mathrm{fft}}} \sum_{\mathbf{k}} \tilde{N}(\mathbf{k})\, e^{i\mathbf{k}\cdot(\mathbf{r}_i - \mathbf{r}_j)} \cdot \sum_{\mathbf{k}'} \mathbf{m}_{\mathbf{k}'}\, e^{i\mathbf{k}'\cdot\mathbf{r}_j} \\
&= -(4\pi M_{\mathrm{s}}) \sum_{\mathbf{k}} \tilde{N}(\mathbf{k})\, e^{i\mathbf{k}\cdot\mathbf{r}_i} \cdot \sum_{\mathbf{k}'} \mathbf{m}_{\mathbf{k}'} \frac{1}{N_{\mathrm{fft}}} \sum_j e^{-i(\mathbf{k}-\mathbf{k}')\cdot\mathbf{r}_j} \\
&= -(4\pi M_{\mathrm{s}}) \sum_{\mathbf{k}} \left[\tilde{N}(\mathbf{k}) \cdot \mathbf{m}_{\mathbf{k}} \right] e^{i\mathbf{k}\cdot\mathbf{r}_i}
\end{aligned}
\tag{3.1}
$$

where $N_{\mathrm{fft}} = L_x \times L_y \times L_z$ is the total number of cells in this regular mesh with periodic boundary conditions (PBC) applied in all three dimensions.

However, a ferromagnetic device studied may just be finite. Suppose we can use a regular mesh with a total scale $N_0 = N_x \times N_y \times N_z$ of cubic or cuboid micromagnetic cells to include or cover the whole shape of the simulated magnetic material or device. In a mesoscopic (nm–μm) device or nano-particle, the regular cells have to be doubled in all three dimensions as with $L_x = 2N_x$, $L_y = 2N_y$ and $L_z = 2N_z$. In a macroscopic (mm or larger) or bulk material, which can be approximated as an infinitely large one, conditions $L_x = N_x$, $L_y = N_y$ and $L_z = N_z$ should be chosen. In a thin film in x–y plane, which has macroscopic scale in-plane and mesoscopic scale perpendicular-to-plane, conditions $L_x = N_x$, $L_y = N_y$ and $L_z = 2N_z$ should be chosen. In all these three cases, the magnetization $\hat{m}_j$ should be set as zero in cells outside the device within the range $N_0 = N_x \times N_y \times N_z$, and in suppositional cells $j_1 > N_x$, $j_2 > N_y$, $j_3 > N_z$, to avoid the miscalculation of $\mathbf{r}_{ij} = \mathbf{r}_i - \mathbf{r}_j$.

Furthermore, the micromagnetic cell has a finite-size (D_x, D_y, D_z) but is not a mathematical point, and thus the PBC should be carefully treated in the model. In a spacially periodic mesh with a total size $N_{\mathrm{fft}} = L_x \times L_y \times L_z$, the demagnetizing matrix $\tilde{N}_{ij} = \tilde{N}(\mathbf{r}_i, \mathbf{r}_j)$ between two cuboid cells has to be calculated as illustrated in Fig. 3.1, where the jth cell is fixed as the $(1, 1, 1)$ cell at origin. The PBC are applied in all three dimensions for displacement $\mathbf{r}_{ij} = \mathbf{r}_i - \mathbf{r}_j$ in the calculation of $\tilde{N}_{ij}$, and thus some cells should be transformed accordingly:

$$
\mathbf{r}_i - \mathbf{r}_j \rightarrow
\begin{cases}
\mathbf{r}_i - \mathbf{r}_j - L_x D_x \hat{e}_1, & \text{if } i_1 - 1 \geq L_x/2 \\
\mathbf{r}_i - \mathbf{r}_j - L_y D_y \hat{e}_2, & \text{if } i_2 - 1 \geq L_y/2 \\
\mathbf{r}_i - \mathbf{r}_j - L_z D_z \hat{e}_3, & \text{if } i_3 - 1 \geq L_z/2
\end{cases}
\tag{3.2}
$$

The ith cuboid cell (where index i means (i_1, i_2, i_3) in 3-D) has six surfaces, which will all contribute to $\tilde{N}_{ij}$. However, since the jth cell is fixed at $(1, 1, 1)$, in the ith

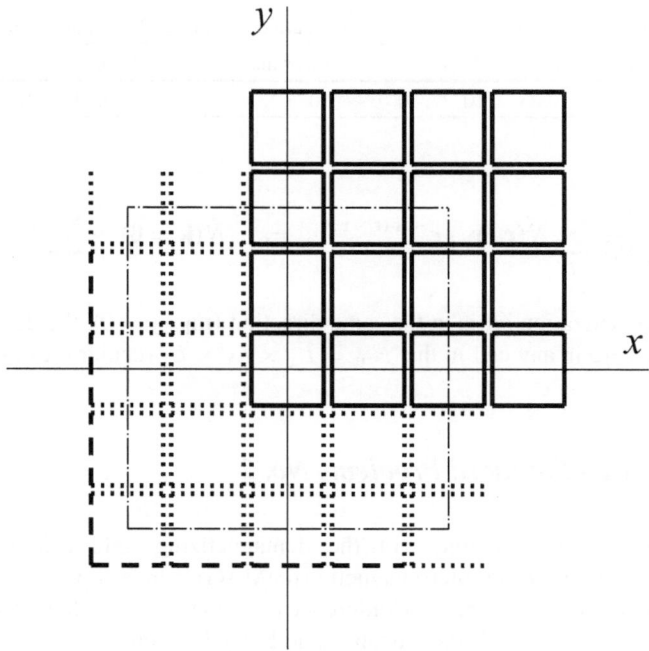

Fig. 3.1 Periodic boundary conditions (PBC) when using FFT method in a regular mesh with finite-size cells: *solid-line boxes* are the $L_x \times L_y$ cuboid cells; the large box of *dash-dotted lines* labels the range w.r.t. the (1, 1) cell where PBC should be applied (outside) or not applied (inside); *dotted lines boxes* are the suppositional micromagnetic cells moved by $-L_x D_x$ horizontally or $-L_y D_y$ vertically; the *dashed lines* mark the special surfaces which should not be calculated together with other five surfaces in the same cuboid cell, but should be considered separately; © [2011] IEEE. Reprinted, with permission, from Ref. [5]

cell at a location $i_1 - 1 = L_x/2$, $i_2 - 1 = L_y/2$ or $i_3 - 1 = L_z/2$, the center of one of the six surfaces is closer to the origin than the criteria $L_x D_x/2$, $L_y D_y/2$ or $L_z D_z/2$ marked by the dash-dotted thin line large box in Fig. 3.1; therefore, these surfaces should not be translated by $L_x D_x$, $L_y D_y$ or $L_z D_z$ to a new position, as marked by the dashed lines. When the demagnetizing matrix $\tilde{N}_{ij}$ is calculated, after the whole cell at the location $i_1 - 1 = L_x/2$, $i_2 - 1 = L_y/2$ or $i_3 - 1 = L_z/2$ was translated to a new position following Eq. (3.2), the contributions to the demagnetizing matrix from the special surfaces marked by dashed lines should be subtracted, and added back to $\tilde{N}_{ij}$ with these surfaces located at original places (marked by solid lines) inside the large box of dash-dotted thin lines before applying Eq. (3.2).

It should be emphasized that this discussion is independent of the choice of the fixed jth cell, because the demagnetizing matrix $\tilde{N}_{ij} = \tilde{N}(\mathbf{r}_i, \mathbf{r}_j)$ only depends on the displacement vector $\mathbf{r}_i - \mathbf{r}_j$. After applying the treatment of demagnetizing matrix calculation discussed in this subsection, we will have a sum rule:

Table 3.3 Simulation parameters in muMAG standard problem No. 2 (assume $M_s = 798$ emu/cc)

d (nm)	D (nm)	α	dt (s)	Error max$\{|d\hat{m}|\}$	A^* (erg/cm)	l_{ex}^N (nm)
25	2.5, 1.25	0.05	$10^{-14}, 10^{-13}$	10^{-6}	25-0.004 $\times 10^{-6}$	25-1

$$\tilde{N}(\mathbf{k}) = \frac{1}{\sqrt{N_{\text{fft}}}} \sum_i \tilde{N}(\mathbf{r}_i, \mathbf{r}_j) \, e^{-i\mathbf{k}\cdot(\mathbf{r}_i - \mathbf{r}_j)} \quad \Rightarrow \quad \tilde{N}(\mathbf{k} = 0) \sim \sum_i \tilde{N}(\mathbf{r}_i, \mathbf{r}_j) = 0$$

(3.3)

when all magnetization $\hat{m}_i$ is in the same direction ($\mathbf{m}_{\mathbf{k}\neq 0} = 0$), the demagnetizing field will be zero in any cell in this $N_{\text{fft}} = L_x \times L_y \times L_z$ regular mesh with PBC.

3.1.2 muMAG Standard Problems No. 2

We can check the correctness of the demagnetizing field calculation using FDM–FFT method by the micromagnetic (muMAG) standard problems No. 2. In the IEEE International Magnetics Conference (Intermag 1994) held in Albuquerque, USA, Robert McMichael, Mike Donahue and Larry Bennett from NIST brought up the proposal of muMAG standard problems [6]. Donahue and Donald Porter began to develop the OOMMF public micromagnetic code after this meeting.

Standard problems No. 2 was suggested by Tom Koehler from IBM Almaden, and proposed by H. Neal Bertram, Alfred Liu, and Chris Seberino from Center for Magnetic Recording Research (CMRR), University of California at San Diego (UCSD). In a cuboid device with length L, width d and t, with relations $L/d = 5$ and $t/d = 0.1$, the reports are requested for the magnetic switching properties with an external field applied on the (111) direction of the cuboid device, and the size d should be scaled with the Néel exchange length $l_{ex}^N = \sqrt{A^*/2\pi M_s^2}$ in a wide range. There is no crystalline anisotropy involved in the problem, thus this is an excellent problem to test the accuracy of the demagnetizing field calculation (Fig. 3.2).

In the standard problem No. 2, all the parameters are scaled. If we let $M_s = 798$ emu/cc, most scaled parameters will have a physical unit, which is easier to understand, as listed in Table 3.3. In the quasi-static M–H loop calculation, one value is set for the Landau damping constant α. Two different meshes with cell size D are used to check the stability of calculation. When $d/l_{ex}^N = 1$, the exchange constant $A^* = 25 \times 10^{-6}$ erg/cm, which is very large, the time interval dt is 10^{-14} s for this special case; in all other cases of A^*, $dt = 10^{-13}$ s is good enough.

Our solutions of standard problem No. 2 agree with most of submitted results in [6], except that a limit of the ratio D/l_{ex}^N is found for dependable results. When the micromagnetic cell size D is larger than the limit $D/l_{ex}^0 \sim 1.25$, the coercivity is not dependable. This is related to the stability of the domain calculation with different cell size in micromagnetics, which will be discussed further in Chap. 4.

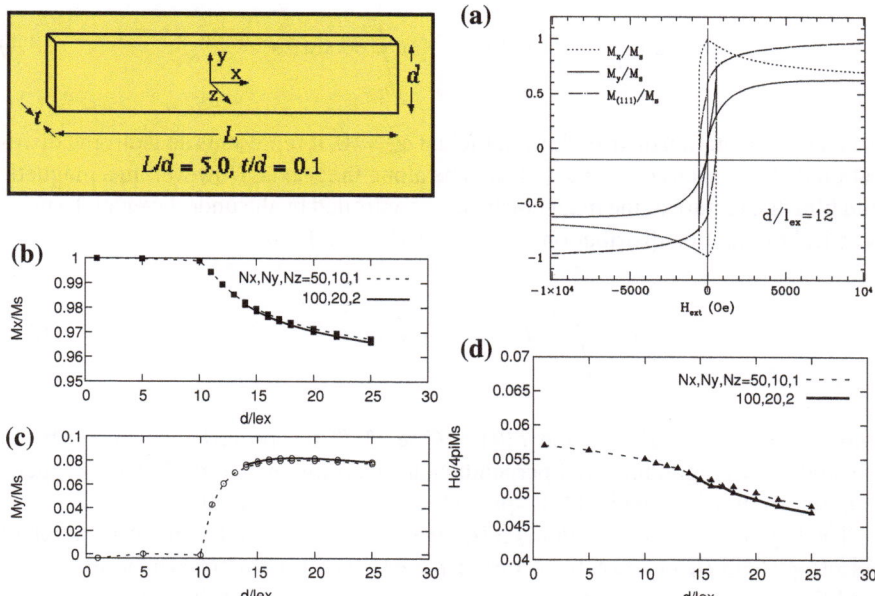

Fig. 3.2 muMAG standard problem No. 2 with two cell sizes $D = d/10$ and $D = d/20$. **a** Calculated hysteresis loops with $d/l_{ex}^N = 12$; *dotted line* for M_x, *solid line* for M_y and *dash-dotted line* for $\langle \mathbf{M} \cdot \mathbf{H}_{ext} \rangle / H_{ext}$ in the conventional M–H loop; **b** remanence along x-axis or the length L; **c** remanence along y-axis or the width d; **d** coercivity of the M–H loop

3.1.3 Anisotropy Field's Orientation and Magnitude Distribution

Practical magnetic recording materials are all alloys, or even composite materials of metals and oxides. Thus there must be substitutional disorder at the atomic scale. Furthermore, defects exist at different scales, even in magnetic grains at nanometer scale there might be twinned structures. Therefore, when we applied the anisotropy energy with cubic symmetry in Eq. (2.54), hexagonal symmetry in Eq. (2.55), or tetragonal symmetry in Eq. (2.56), both the orientations of the anisotropy field's crystal axes $\hat{k}_1, \hat{k}_2, \hat{k}_3, \hat{k}_a, \hat{k}_b, \hat{k}_c$ and the magnitude of the anisotropy field constant $H_k = 2K_1/M_s$ will have distributions.

The choice of orientation distribution for $\mathbf{H}_k$ also depends on the degree of disorder in the crystalline phase. If the magnetic alloy is highly ordered, then we can just chose an orientation distribution $f(\theta)$ for $\hat{k}_3$ in FCC or BCC lattice, or the c-axis $\hat{k}_c$ in HEX or TET lattice; for the other two crystal axes $\hat{k}_1, \hat{k}_2$, or $\hat{k}_a, \hat{k}_b$, we can just set them orthogonal to $\hat{k}_3$ or $\hat{k}_c$, respectively. However, if in reality magnetic alloys have low order, we have to control the distributions of $\hat{k}_a, \hat{k}_b, \hat{k}_c$ independently, and these three crystal axes are even not orthogonal to one another.

In principle, the orientation distribution $f(\theta)$ in 3-D free space should satisfy:

$$1 = 2\pi \int_0^\pi d\theta \sin\theta f(\theta) = C_0 \int_0^\pi d\theta \sin\theta \, e^{-\alpha_\theta \sin^2\theta}. \tag{3.4}$$

When the orientation distribution coefficient $\alpha_\theta = 0$, it represents an isotropic distribution in 3-D; when $\alpha_\theta \to \infty$, $\hat{k}_c$ has to be along the z-axis. However, in a magnetic thin film, the texture of the magnetic layer is controlled by the underlayer, and a more practical orientation distribution is a 2-D distribution for θ:

$$1 = \int_0^\pi d\theta f(\theta) = C_1 \int_0^\pi d\theta \, e^{-\alpha_\theta \sin^2\theta}. \tag{3.5}$$

This 2-D orientation distribution $f(\theta) = C_1 e^{-\alpha_\theta \sin^2\theta}$ is proved to be reasonable in the studies of longitudinal and perpendicular recording media, FeCo soft magnetic thin film for writers, and in TMR spin valve multilayers for readers.

The log-Gaussian distribution $f_0(H_k)$ was usually utilized for the distribution of anisotropy field H_k in magnetic recording media [8]. The author used to brought up an modified log-Gaussian distribution $f_1(H_k)$ for anisotropy fields [9]. The log-normal and the modified log-normal distributions for anisotropy fields are:

$$f_0(H_k) = C_0 \, e^{-\ln^2(H_k/H_k^0)/\beta^2} / H_k; \tag{3.6}$$

$$f_1(H_k) = C_1 \, e^{-\ln^2(H_k/H_k^0)/\beta^2} \, e^{-(H_k/H_k^0)^2}, \tag{3.7}$$

where H_k^0 is the average anisotropy field constant and β is the magnitude distribution coefficient for H_k, which are both inputs of a micromagnetic model.

We tried to understand the magnitude distribution of H_k based on an simple analytical model analogy to the Becker's theory or Stoner–Wohlfarth model, where the twinned or n-fold structure exists in a nanoparticle [7]. In L10-FePt nanoparticles as small as 3.4 nm, the n-fold structures was discovered by high-resolution TEM. The decahedron particle is composed of five twinning parts T1–T5 with the twinning planes being the {111} planes. The L10-FePt's [001] easy axis for each twinning part lies in the plane perpendicular to the center atom column, and it is perpendicular to the edge of the projection pentagon, as shown in the atomic model in Fig. 3.3. In real particles, the volumes of five twinning regions are different.

The model for anisotropy of a nanoparticle is set up based on two assumptions: (1) the magnetizations of all twinning or n-fold parts rotate coherently; (2) the volumes V_i of twinning regions distribute as the log-Gaussian distribution $\exp(-\ln^2(V_i/V_0)/\beta_\zeta^2)/V_i$. If the nanoparticle consists of n-fold structure, the Cartesian coordinate system is established with any one of twin boundaries as x-axis. The free energy of the nanoparticle with n-fold structures can be written as follows:

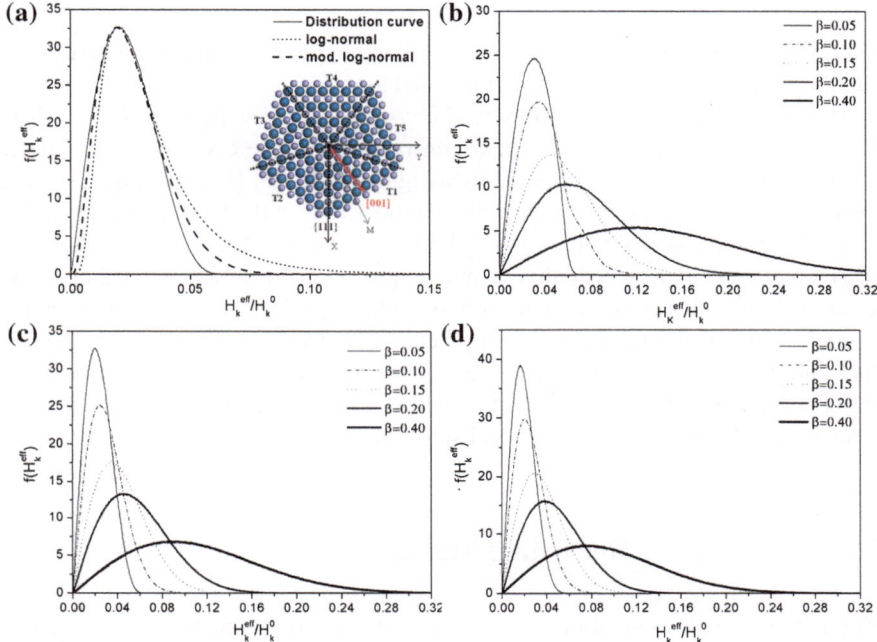

Fig. 3.3 Normalized distribution of effective anisotropy field H_k^{eff} in magnetic nano-particles with twinned or n-fold structures. **a** Curve fittings of the effective anisotropy field distribution $f(H_k^{\text{eff}})$ (where $\beta_\zeta = 0.05$, fivefold structure); **b** threefold structure; **c** fivefold structure; **d** sevenfold structure; © [2009] IEEE. Reprinted, with permission, from Ref. [7]

$$\mathscr{F} = \mathscr{E}_{\text{ext}} + \mathscr{E}_{\text{a}} = -V M_s(\hat{m} \cdot \mathbf{H}_{\text{ext}}) + K \sum_{i=1}^{n} V_i (\hat{k}_i \times \hat{m})^2, \tag{3.8}$$

where $\hat{k}_i$ ($i = 1 - n$) is the unit vector of anisotropy field for each twinning or n-fold part, which is parallel to the direction of the corresponding easy axis [001]. The V is the total volume of the nanoparticle, namely $\sum V_i$. The anisotropy energy $\mathscr{E}_{\text{a}}$ can be derived and written in a form similar to Eqs. (2.70) and (2.71):

$$
\begin{aligned}
\frac{\mathscr{E}_{\text{a}}}{V} &= \frac{K}{V} \sum_{i=1}^{n} V_i \, \sin^2\left(\frac{2i-1}{n}\pi - \theta\right) \\
&= \frac{1}{2}K - \frac{1}{2}K \, (A\cos 2\theta + B\sin 2\theta) \qquad A = \sum_{i=1}^{n} \frac{V_i}{V} \cos\left(\frac{4i-2}{n}\pi\right) \\
&= \frac{1}{2}K - \frac{1}{2}K\sqrt{A^2 + B^2}\cos(2\theta - 2\theta_k) \qquad B = \sum_{i=1}^{n} \frac{V_i}{V} \sin\left(\frac{4i-2}{n}\pi\right) \\
&= C_0 + K\sqrt{A^2 + B^2}\sin^2(\theta - \theta_k) \qquad \theta_k = \frac{1}{2}\arctan\left(\frac{B}{A}\right)
\end{aligned}
\tag{3.9}
$$

where θ is the angle between $\hat{m}$ and x-axis; θ_k is the angle between the whole particle's effective uniaxial anisotropy orientation $\hat{k}$ and the x-axis.

The effective uniaxial anisotropy field constant is $H_k^{\text{eff}} = H_k^c \sqrt{A^2 + B^2}$, where $H_k^c = 2K/M_s$ is the anisotropy field constant in a single crystal phase, and the constants A and B can be computed by assigning volumes $\{V_i\}$ into Eq. (3.9). The constants A, B only depends on the volume ratio Vi/V of the twinned parts; H_k^{eff} is only related to the scaled volumes $\{Vi/V\}$ but not the absolute values of volume V of nanoparticles. In the calculation, a total number of 10^7 tries for twinned or n-fold region volumes $\{V_i\}$ are carried out following the log-Gauss distribution. Then the calculated effective uniaxial anisotropy field H_k^{eff} also has a distribution, and this distribution is closer to the modified log-Gaussian distribution in Eq. (3.7), as shown in Fig. 3.3a. When the number n for twinned structures is larger, the H_k^{eff} is smaller compared to $H_k^c = 2K/M_s$ in the ideal single crystal phase.

3.2 Perpendicular Recording Media

The micromagnetic simulations of hard magnetic recording media were developed since 1980s. In this section, two classes of micromagnetic models are introduced and compared for perpendicular recording media, with different levels of microstructure description. In the first simple model, one magnetic grain is modeled as a micromagnetic cell. In the second model, the polycrystalline structure is simulated, the symmetries in the crystal grains are considered, and the magneto-elastic field related to the underlayer is included. We will see when the models are more realistic, the simulated M–H loops in 3-D space are closer to experiment.

First, we can build a simple micromagnetic model for the CoPtCr–SiO$_2$/Ru/SUL perpendicular media [10], as discussed in Sect. 1.3.1. A cylindrical shaped magnetic grain is chosen as the micromagnetic cell, as seen in Fig. 3.4. The scaled effective magnetic field $\mathbf{h}_i = \mathbf{H}_{\text{eff}}/H_k^a$, where H_k^a is the averaged anisotropy field of log-normal distribution in Eq. (3.6), can be expressed as:

$$\mathbf{h}_i = h_{\text{ext}} + h_k^s \, (\hat{e}_y \cdot \hat{m}_i) \, \hat{e}_y + h_k^i \, (\hat{k}_i \cdot \hat{m}_i) \, \hat{k}_i$$

$$+ h_{\text{ex}} \sum_j^{NN} (\hat{m}_j - \hat{m}_i) - h_m \sum_{\mathbf{k}} \left[\tilde{N}(\mathbf{k}) \cdot \mathbf{m_k} \right] e^{i\mathbf{k} \cdot \mathbf{r}_i} \qquad (3.10)$$

where the scaled external field $h_{\text{ext}} = H_{\text{ext}}/H_k^a$, the scaled magnetostatic interaction field $h_m = 4\pi M_s/H_k^a$, the scaled exchange field $h_{\text{ex}} = 2A^*/(H_k^a M_s a_L^2) = A^*/(K_1^a a_L^2)$, and the scaled shape anisotropy field constant $h_k^s = h_m[\tilde{N}_{11}^s(0,0) - \tilde{N}_{22}^s(0,0)]$. The demagnetizing matrix of a cylinder can be calculated as:

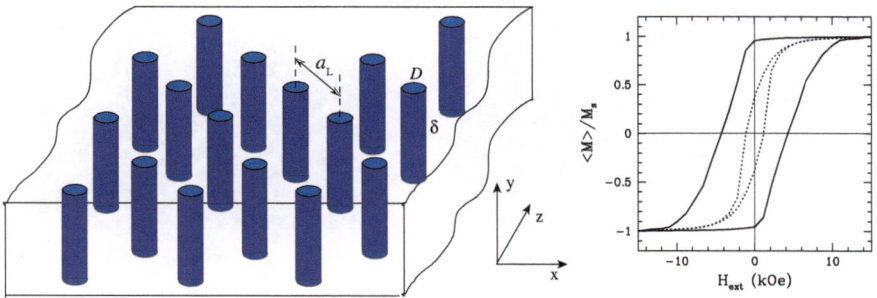

Fig. 3.4 Simple micromagnetic model for CoCrPt–SiO$_2$ and simulated M–H loops (*solid line* is for perpendicular loop and *dotted line* is for longitudinal loop)

Table 3.4 Simulation parameters in a simple model of CoCrPt–SiO$_2$

a_L (nm)	D (nm)	δ (nm)	M_s (emu/cc)	H_k^a (kOe)	α_θ	A^* (erg/cm)
8	7	12	560	11	8.8	0.28×10^{-6}

$$\tilde{N}(\mathbf{r},0) = -\frac{1}{4\pi} \iint_S d^2\mathbf{r}' \frac{(\mathbf{r}-\mathbf{r}')\hat{n}'}{|\mathbf{r}-\mathbf{r}'|^3} = -(\tilde{u}_+ + \tilde{u}_- + \tilde{u}_c) \qquad (3.11)$$

where the contributions from the up and bottom surfaces $\tilde{u}_\pm$ have similar forms ($\hat{n}' = \pm\hat{e}_z$, the integration over ρ can be done analytically), the contribution from the side surface $\tilde{u}_c$ is more important if the size of a cell $\delta > D$ ($\hat{n}' = \cos\theta\hat{e}_x + \sin\theta\hat{e}_y$, the integration over z can be done analytically).

The parameters in the simulation as listed in Table 3.4 is chosen close to the experiment of [10]. The simulated loops are shown in Fig. 3.4. The perpendicular loop is close to the experiment [10], except that the tail is longer, due to the usage of the log-normal distribution but not the modified log-normal distribution for h_k^i. However, the longitudinal loop in plane is quite different from measurement, this reveals the weakness of the microstructure description in this simple model.

The polycrystalline microstructure was included in the micromagnetic model in a study of FePt media with very high anisotropy in 2006 [11]. In the well ordered FePt L10-phase, the exchange length $l_{ex}^B = \sqrt{A^*/K_1}$ is as small as 1–5 nm, which is much less than the grain size. Therefore the unit of simulation has to be a grid instead of a magnetic grain. The magnetic grains were created by a random initial nucleation and the grain-growth simulation afterwards, as seen in Fig. 3.5a; in [11] the grain boundary is wider. The magnetization at grain boundary is set as zero. The effective field expression was almost the same as Eq. (3.10) except that two exchange terms (h_{ex}^1 is intra-grain exchange between neighboring grids and h_{ex}^2 is inter-grain exchange across the grain boundary) are introduced:

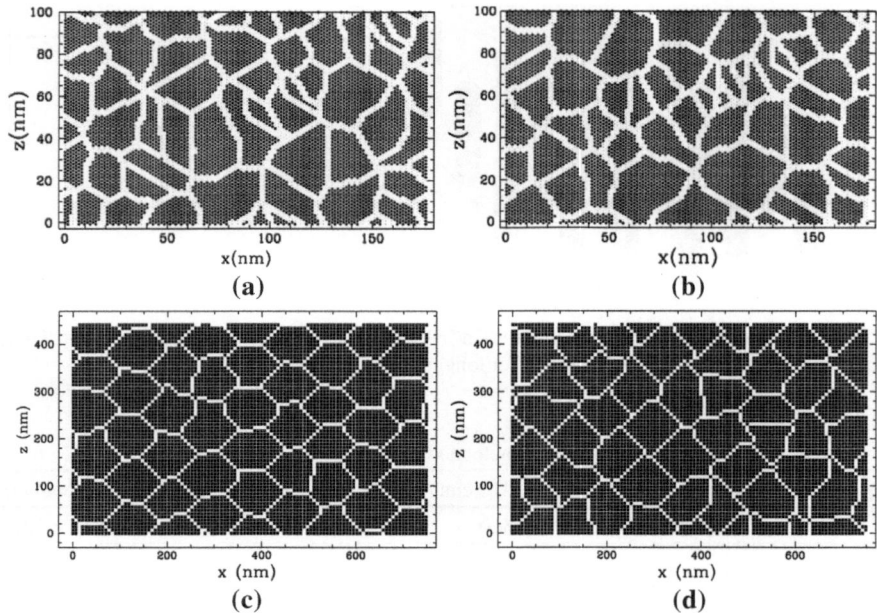

Fig. 3.5 Grain-growth simulation on a regular mesh. **a** Random seeds, hexagonal mesh; **b** triangular-arranged seeds, 10 random move, hexagonal mesh; **c** triangular-arranged seeds, 1 random move, cuboid mesh; **d** triangular-arranged seeds, 10 random move, cuboid mesh

$$\mathbf{h}_i = h_{\text{ext}} + h_{\text{k}}^{\text{s}} \, (\hat{e}_y \cdot \hat{m}_i) \, \hat{e}_y \; + h_{\text{k}}^i \, (\hat{k}_i \cdot \hat{m}_i) \, \hat{k}_i \; + h_{\text{ex}}^1 \overset{\text{intra}}{\underset{j}{\sum}} (\hat{m}_j - \hat{m}_i)$$

$$+ h_{\text{ex}}^2 \overset{\text{inter}}{\underset{j}{\sum}} (\hat{m}_j - \hat{m}_i) - h_{\text{m}} \sum_{\mathbf{k}} \left[\tilde{N}(\mathbf{k}) \cdot \mathbf{m_k} \right] e^{i\mathbf{k}\cdot\mathbf{r}_i} \qquad (3.12)$$

The simulation longitudinal and perpendicular loops agreed with experiment [12].

The grain size distribution can be controlled if initially the seeds of grain-growth are triangular-arranged on the regular mesh, then let the seeds do the random walk to neighbor sites. More steps of random move result in a wider grain size distribution. If the regular mesh of micromagnetic cells has hexagonal symmetry, actually the Voronoi structure of polycrystalline thin film is easier to be achieved, as seen in Fig. 3.5b. However, the FFT on a hexagonal or triangular mesh has some problems. In Eq. (3.1) for FFT of demagnetizing field, if we use a set of primitive vectors **a**, **b**, **c** for the hexagonal mesh, the FFT can be done as:

$$\mathbf{H}_{\text{m}}(I, J, K) = -(4\pi M_{\text{s}}) \sum_{m,n,l} \left[\tilde{N}(m, n, l) \cdot \mathbf{m}(m, n, l) \right] e^{i2\pi(Im+Jn+Kl)}$$

$$\mathbf{r} = I\mathbf{a} + J\mathbf{b} + K\mathbf{c}; \qquad \mathbf{k} = m\mathbf{a}^* + n\mathbf{b}^* + l\mathbf{c}^* \quad \text{(reciprocal lattice)}. \quad (3.13)$$

Table 3.5 Simulation parameters in micromagnetic model of FePt/Pt/CrW

D_x, D_z (nm)	D_y (nm)	D_g (nm)	δ (nm)	M_s (emu/cc)	K_{u1} (erg/cc)	H_k (kOe)
7	9	45	18	700	5.26×10^6	15
α_θ (c, a)	β	A_1^* (erg/cm)	A_2^* (erg/cm)	K_{u2}/K_{u1}	K_c/K_1	H_σ (kOe)
4, 2	0.1	2.0×10^{-6}	0.9×10^{-6}	0	4	15

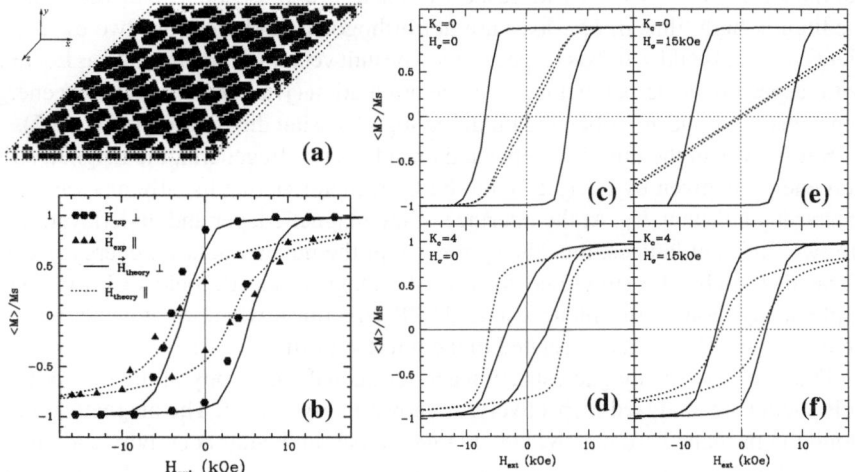

Fig. 3.6 Simulation of FePt media. **a** 3-D view of microstructure; **b** simulated perpendicular and in-plane loops compared with experiment; **c–f** effects of tetragonal anisotropy term $K_c = 0, 4K_{u1}$ and magnetostriction field $H_\sigma = 0, 15\,\text{kOe}$ on the M–H loops

However, the treatment of periodic boundary condition in Fig. 3.1 is hard to achieve on a triangular lattice. Thus we chose to use cuboid mesh (with cubic, tetragonal or orthorhombic symmetry) instead. In Fig. 3.5c, d, the polycrystalline microstructure with different grain size distribution are achieved respectively.

A 3-D micromagnetic model is built up to analyze the relationship between the magnetic property and the microstructure of FePt/Pt/CrW thin films, with careful discussions of the tetragonal symmetry and the magnetostriction [13]. The geometrical and magnetic parameters of the simulated medium are listed in Table 3.5, which correspond to the experiments in [14]. In the beginning of the simulation, a tetragonal mesh is created with a lattice constant D_x, D_y, D_z on the order of Bloch exchange length l_{ex}^B. The simulated polycrystalline thin film is plotted in Fig. 3.6a. The averaged magnetic grain size $D_g = 45\,\text{nm}$ and the film thickness $\delta = 18\,\text{nm}$ are both larger than grid size, which match the microstructure in [14].

The micromagnetic simulation unit is the cell in the regular mesh. The effective magnetic field has been given in Eqs. (2.57)–(2.61) in Chap. 2, just note that the position vectors $\mathbf{r}$ and $\mathbf{r}'$ are discretized on the tetragonal mesh. In the hard magnetic layer, the magnetization M_s at the grain boundary is set as zero, which is true for recording media with well grain–grain segregation. Two exchange constants are introduced: A_1^* is for exchange between two neighbor grids inside a grain and A_2^* is for exchange between two neighbor grids across the grain boundary.

The crystalline anisotropy energy density of the L10-FePt with tetragonal symmetry has been given in Eq. (2.56), where K_{u1} and K_{u2} are the first and second order of the uniaxial anisotropy energy around the c-axis $\hat{k}_c$, and K_c are the tetragonal anisotropy energy term with respect to the a, b-axis $\hat{k}_a$, $\hat{k}_b$. The disorder in this medium is high, thus a, b, c-axes are not orthogonal, and there are two α_θ coefficients for c-axis and a, b-axis respectively. The unit vector $\hat{k}_c$ is distributed as Eq. (3.5) with respect to the perpendicular-to-plane orientation (y-axis); the anisotropy energy constants, K_{u1} and K_c, obey the modified log-Gaussian distribution in Eq. (3.7).

Stress exits in the thin film deposited on substrates. In general, the magnetostriction energy is given by Eq. (2.52), but bulk magnetostriction usually just varies the anisotropy constant K_1. At the interface of the magnetic layer and underlayer, there are two different lattices, thus the symmetry of the magneto-elastic energy is complicated. Usually the simplified form of the magneto-elastic field in Eq. (2.61) is applied, say along an in-plane x-axis ([100] direction), where the magneto-elastic field constant $H_\sigma = H_{me}$ is related to the interfacial stress (2.61).

The effects of the magnetostriction and tetragonal anisotropy of the L10-ordered FePt media are analyzed respectively, as shown in Fig. 3.6c–f. The magnetostriction along the in-plane hard axis would enlarge the perpendicular coercivity and largely augment the saturation field of the longitudinal loop along the x-axis. The tetragonal anisotropy, especially a large in-plane anisotropy energy $K_c = 4K_{u1}$, tends to enlarge the longitudinal coercivity, decrease the perpendicular coercivity, squareness and coercive squareness, and conduce the formation of the open-up between the tails of the longitudinal and perpendicular loops. Therefore the experimental M–H loops [14] could be explained, as shown in Fig. 3.6b, f.

The magnetic properties of the FePt thin films shown in Fig. 3.6 and Table 3.5 can not be used as perpendicular recording media, since their grain size D_g, the tetragonal anisotropy K_c, and the cross-boundary exchange A_2^* are too large. Actually the current recording media are usually CoPt-oxide media with a relatively soft Co-oxide cap layer, which is the Exchange-Coupled-Composite (ECC) media [15] with both high thermal stability and suitable coercivity for writability. We will give a simulation for the hard magnetic CoPt–TiO$_2$ layer here (Table 3.6).

A 3-D micromagnetic model is built up to analyze the CoPt–TiO$_2$ hard layer. The simulated polycrystalline thin film is plotted in Fig. 3.7a, with an averaged magnetic grain size $D_g = 7\,nm$ and the film thickness $\delta = 16\,nm$, which match the microstructure in ECC media's experiment (from Dr. Ying Wang's Ph.D thesis in Lanzhou University, P. R. China). In the hard layer, the magnetization M_s at the grain boundary ($\leq 2\,nm$) is still set as zero. The crystalline anisotropy of CoPt with hexagonal symmetry has been given in Eq. (2.55), where K_1 is the first order of the uniaxial anisotropy energy around the c-axis. The orientation distribution coefficient α_θ is not large, maybe due to the substitutional disorder of Co-Pt alloy. The exchange constant A_1^* between two neighbor cells inside a grain is smaller than the value for Fe, Co, Ni metal alloys, due to the influence of oxide co-sputtering, however the

Table 3.6 Simulation parameters of CoPt–TiO$_2$ hard layer in ECC media

D (nm)	D_g (nm)	δ (nm)	M_s (emu/cc)	K_1 (erg/cc)	H_k (kOe)
2	7	16	619	4.95×10^6	16
α_θ (c)	β	A_1^* (erg/cm)	A_2^* (erg/cm)	H_{ex}^1 (kOe)	H_σ (kOe)
1	0.1	0.2×10^{-6}	0.01×10^{-6}	16	10

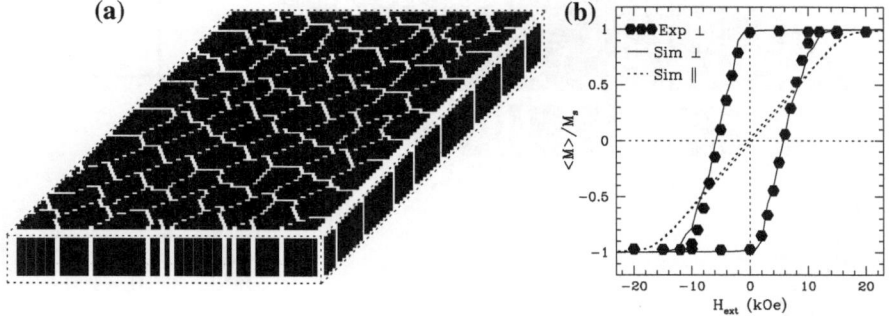

Fig. 3.7 Simulation of CoPt–TiO$_2$ hard layer in ECC media. **a** Microstructure; **b** M–H loops

respective exchange field constant H_{ex}^1 is still in the same order as H_k, which insures the uniform rotation of a grain. The cross-boundary exchange constant A_2^* is very small, which means the grain segregation is well done. However, practical magnetic recording media need higher ordering in the crystal phase, i.e., better orientation of H_k (larger α_θ) and larger intra-grain exchange constant A_1^*.

3.3 Soft Magnetic FeCo Thin Film

FeCo is the magnetic material with highest magnetization; therefore in write heads of hard disk drives, it is chosen as the main pole materials to provide higher write field. In 1960, Hall measured the anisotropy and magnetostriction coefficient of FeCo alloy [16]. When the atomic ratio x in Fe$_{1-x}$Co$_x$ is less than 40%, the structure is still BCC and the easy axis is $\langle 100 \rangle$, the same as pure Fe. The Fe$_{65}$Co$_{35}$ alloy with highest room temperature saturation $4\pi M_s = 2.45$ T, which is often used in writers, has an anisotropy energy constant $2K_1 = 1.0 \times 10^5$ erg/cm^3 in Eq. (2.54), and the respective anisotropy field constant $H_k^0 = 4K_1/M_s = 102$ Oe. The magnetic properties of FeCo thin films depend on the substrates. If FeCo is deposited on glass, the in-plane M–H loops are isotropic and the coercivity is in the range of 50–120 Oe, which is on the order of H_k^0. If FeCo is deposited on polycrystalline metallic substrates such as Co, Cu, NiFe, Ru, Fe and NiCr [17], the in-plane coecivity in the easy axis can be decreased to 5–20 Oe, and the coercivity in the hard axis is on the order of 1 Oe; thus FeCo thin film has a nice in-plane anisotropy.

(a)

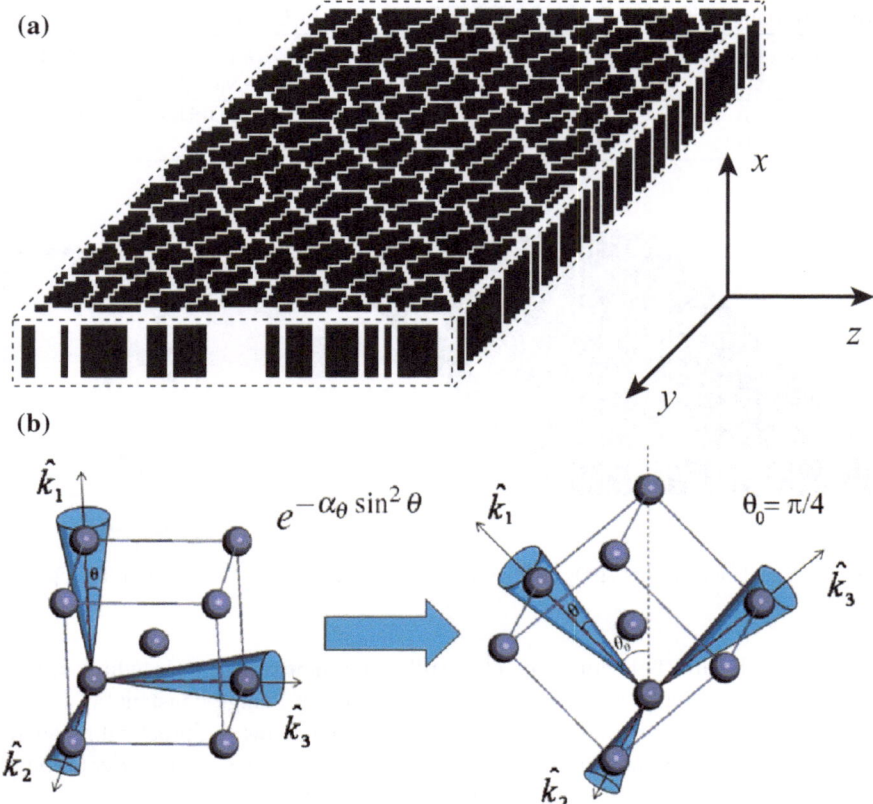

(b)

Fig. 3.8 Simulation of microstructure in FeCo thin film. **a** Polycrystalline thin film; **b** texture

The theory or simulation of soft magnetic thin film is a hard topic in applied magnetism, although the loops of soft irons had been measured 150 years ago. The coercivity of FeCo in the easy axis is on the order of 10 Oe, which is smaller than the anisotropy field $H_k^0 \sim 100$ Oe, and much smaller than the demagnetizing field and the exchange field ($\sim 10,000$ Oe) among micromagnetic cells.

The author introduced a micromagnetic model for FeCo based on the microstructure of a metallic thin film [18], which is in the same frame as the model of hard magnetic storage thin films. The polycrystalline thin film is simulated based on a regular mesh of $2 \times 2 \times 2$ nm cubic micromagnetic cells. The film thickness δ is 16 or 32 nm, the grain size is 10 nm and the grain boundary width is less than 2 nm, as seen in Fig. 3.8a. PBC are applied in the y–z plane. The simulation parameters are treated separately for the crystal phase in the grain and the amorphous phase at grain boundary, as listed in Table 3.7.

It has been known for a long time that many amorphous alloys of Fe are soft magnetic. The amorphous phase at the grain boundary of FeCo polycrystalline thin

Table 3.7 Simulation parameters of FeCo soft magnetic layer

Micromagnetic cell location	$4\pi M_s$ (T)	$2K_1$ (erg/cm^3)	α_θ	H_e (Oe)	H_k^0 (Oe)	H_σ (Oe)
Crystal grain	2.4	5.0×10^4	10	20,000	52.5	$\sim$200
Grain boundary	95% $\times 2.4$	4.0×10^4	1	10% $\times 20,000$	45.0	$\sim$200

film is also ferromagnetic, because in experiment the total saturation of the thin film is not far from the single crystal value $4\pi M_s = 2.4$ T. In Table 3.7, the saturation at grain boundary $M_s^a = 95\% M_s$ is one of the crucial parameters to achieve a nice in-plane anisotropy; when $M_s^a = 90\% M_s$ the coercivity in the hard axis will increase evidently, as seen in Fig. 3.9. The atoms are arranged randomly at grain boundary, thus the exchange field constant H_e^a related to a micromagnetic cell in the amorphous phase is set as only 10% of the value in the crystal phase.

At room temperature, $Fe_{65}Co_{35}$ has a BCC structure with a (110) texture. Therefore the anisotropy energy is described by Eq. (2.54) with a cubic symmetry. To achieve the (110) texture, the orthogonal cubic axes $\hat{k}_1$, $\hat{k}_2$, $\hat{k}_3$ rotate following the processes in Fig. 3.8b. First the $\hat{k}_1$ axis rotate from x-axis by angle θ following a distribution $f(\theta) = \exp(-\alpha_\theta \sin^2 \theta)$ with a random ϕ angle for $\hat{k}_2$ and $\hat{k}_3$ axes. Then the $\hat{k}_1$, $\hat{k}_2$, $\hat{k}_3$ rotate around the y-axis by $\theta_0 = 45°$ to achieve the (110) texture. The simulation result does not change much if the magnitude distribution of H_k in Eq. (3.7) is included. The anisotropy constant K_1^a at amorphous grain boundary is smaller than that in a crystal grain; actually the result is not sensitive with K_1^a: the soft magnetic property is good with $2K_1^a$ from 0 to 4×10^4 erg/cm^3.

The most important parameter to explain the in-plain anisotropy is the magneto-elastic field related to the underlayer. Berkowitz used to point out that the exchange at the interface of the magnetic layer and the ferromagnetic or antiferromagnetic underlayer will cause anisotropy [19]. Nakamura group studied FeCo/NiFe(Cr) thin films and declared that the improvement of the in-plain anisotropy is related to the lattice deformation caused by the underlayer [20]. The stress between the magnetic layer and underlayer is originated from the thermal stress (different thermal expansion coefficients) and the intrinsic stress (different lattice constants). Thus the stress must be distributed near the interface; in this model of FeCo, a magneto-elastic field in Eq. (2.61) is introduced just in the micromagnetic cells adjacent to the underlayer:

$$\mathbf{H}_\sigma = H_{me} m_z \hat{e}_z + H'_{me}(m_y \hat{e}_z + m_z \hat{e}_y) \simeq H_\sigma m_z \hat{e}_z. \qquad (3.14)$$

The saturation magnetostriction coefficient in bulk FeCo is very high: $\lambda_s = (47 \pm 4) \times 10^{-6}$ [21]. The stress between the FeCo layer and underlayer is on the order of GPa [22]; if we use the bulk value of λ_s, the effective magneto-elastic field $H_\sigma \sim 3\lambda_s \sigma / M_s = 738$ Oe, which is too large compared to the estimation of the in-plane anisotropy in the FeCo thin film. Actually in FeCo thin film, the measured value of λ due to interfacial stress is much smaller than the bulk value λ_s; and the magneto-elastic constant H_σ is chosen as 200 or 300 Oe here.

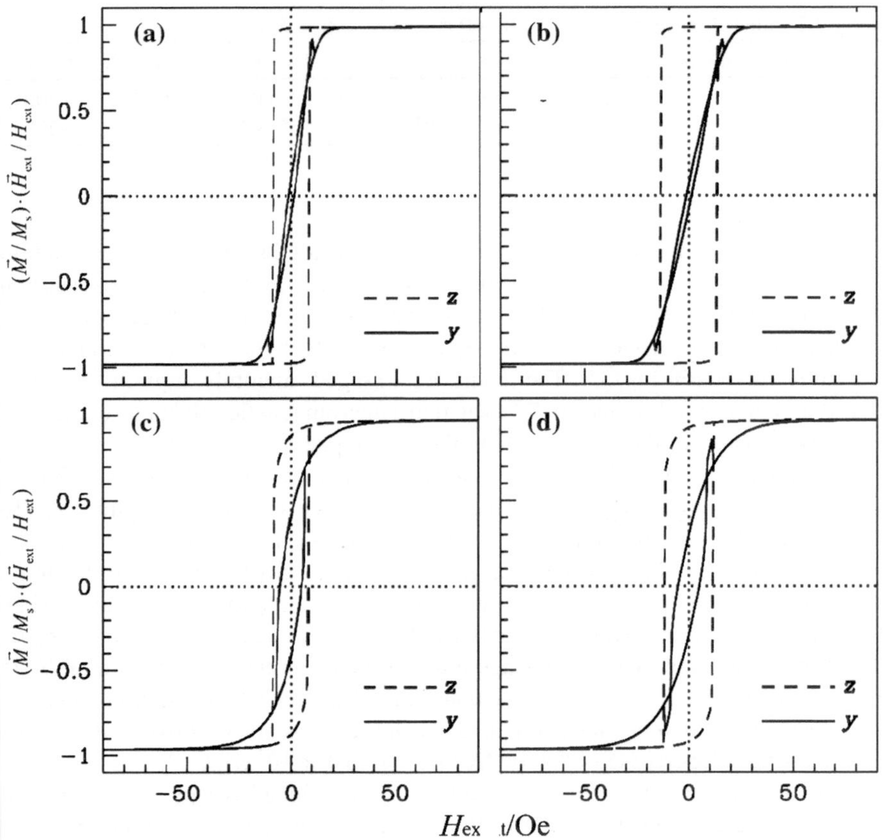

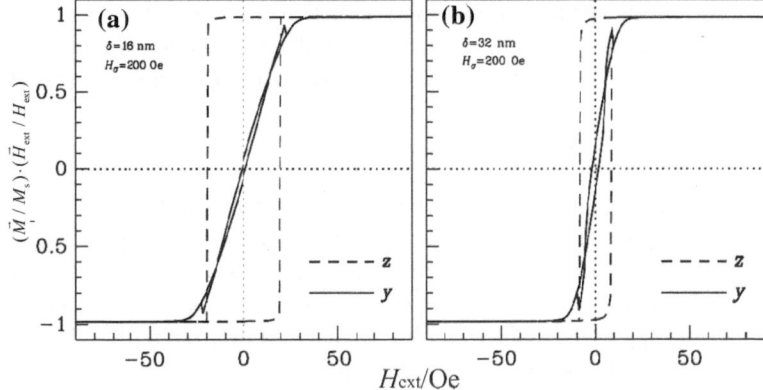

Fig. 3.9 Simulated M–H loops of FeCo layer with thickness $\delta = 32\,\text{nm}$. **a** Magneto-elastic field $H_\sigma = 200\,\text{Oe}$, saturation ratio $M_s^a/M_s = 95\%$; © [2010] IEEE. Reprinted, with permission, from Ref. [18]; **b** $H_\sigma = 300\,\text{Oe}$, $M_s^a/M_s = 95\%$; **c** $H_\sigma = 200\,\text{Oe}$, $M_s^a/M_s = 90\%$; **d** $H_\sigma = 300\,\text{Oe}$, $M_s^a/M_s = 90\%$

Fig. 3.10 **a** M–H loops in a thinner FeCo layer; **b** M–H loops with regular-arranged grains

The calculated in-plane M–H loops are plotted in Fig. 3.9 for a 32 nm FeCo layer, when $H_\sigma = 200$ Oe, the easy axis coercivity H_{ce} is 9 Oe and the hard axis coercivity H_{ch} is 2 Oe; when $H_\sigma = 300$ Oe, $H_{ce} = 14$ Oe and $H_{ch} = 2$ Oe. When the saturation at grain boundary M_s^a decreases to $90\% M_s$, the shapes of loops are different. In a 16 nm layer with $H_\sigma = 200$ Oe and $M_s^a = 95\% M_s$, H_{ce} is larger (20 Oe), because the effect of magneto-elastic field applied on cells adjacent to underlayer is relatively stronger. The spike in the hard-axis loop can be seen in experiment [20], which might be due to the regular arrangement of grains, as compared in Figs. 3.9a and 3.10b.

3.4 Tunneling Magnetoresistive Spin Valve

TMR junctions with exchange bias structure are extensively employed in devices such as magnetic read heads in hard disk drives, memory cells in magnetoresistive random access memory and magnetic sensors. In magnetic recording industry, MR heads, GMR heads and TMR heads were developed in 1991, 1994 and 2004, respectively. Almost all perpendicular recording hard disk drives use TMR readers after 2005.

The previous theories of GMR or TMR are basically spin-dependent electronic theory. In 1954, Ruderman and Kittel found that two ions in the Bloch electrons gas have effective exchange energy, called RKKY interaction later [23], which is the key to explain the large MR ratio in GMR multilayers. In 2004, Zhang and Butler used the layer Korringa–Kohn–Rostoker (KKR) first principle calculation technique and predicted the huge MR ratio (6,000%) of FeCo/MgO/FeCo, which is the key electronic theory of TMR multilayers [24]. However, the spin-dependent electronic theory just explains the MR ratio with parallel/antiparallel atomic spins in the free layer (FL) and pinned layer (PL), but it cannot explain the shape of the measured R–H loops or M–H loops of TMR multilayers.

In 2004, influenced by Zhang and Butler's theoretical prediction, TMR multilayers evolved from "AlO" age to "MgO" era. In this section, we would like to present the micromagnetic theory of the textured TaN/IrMn/CoFe(PL)/MgO/CoFe(FL)/TaN spin valve with a MR $\sim 150\%$ developed by Parkin's group in IBM [25], and this is the core structure used in TMR head after 2005.

In the micromagnetic model of TMR spin valve, the pinned CoFe layer is composed of 0.8 nm thick $Co_{84}Fe_{16}$ and 3 nm thick $Co_{70}Fe_{30}$, and the free CoFe layer is a 15 nm thick $Co_{84}Fe_{16}$ [25]. The micromagnetic cell is $1.5 \times 1.5 \times 1.5$ nm^3, where the cell size $D = 1.5$ nm is smaller than the Néel exchange length l_{ex}^N, which is about 2.66 nm for the PL and 2.85 nm for the FL, respectively. The thickness of PL is chosen as 3 nm in the model for convenience. The thickness of FL is still 15 nm, the same as experiment. The MgO layer thickness is set as 3 nm, while in experiment it is either 3.1 or 2.9 nm. Therefore, along perpendicular z-axis, there are two cells for the PL, two cells for the MgO barrier and ten cells for the FL. The total size of the regular mesh is $32 \times 32 \times 14$ micromagnetic cells. In the M–H and R–H loop calculation, PBC are applied in-plane.

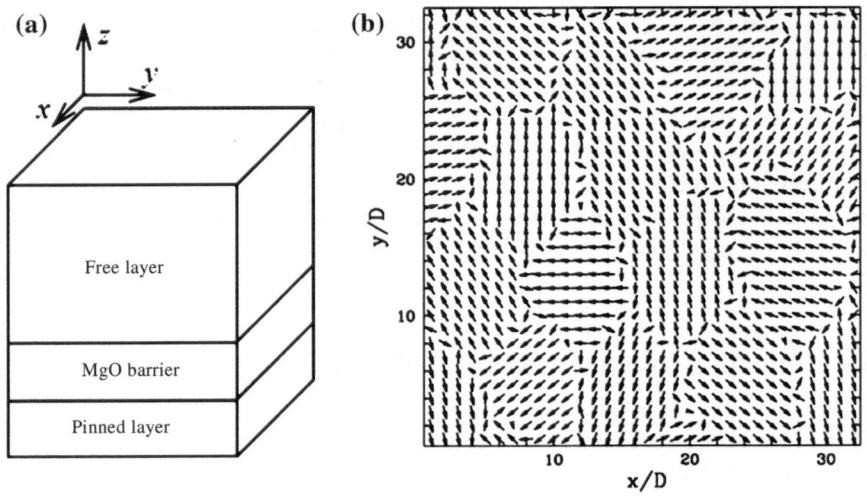

Fig. 3.11 **a** Scheme of TMR spin valve; **b** for each cell in a film plane with (100) texture, projection of [001] directions or $\hat{k}_3$ axes of the cubic anisotropy in a crystal grain with a BCC symmetry, and an amorphous phase at the grain boundary where $\hat{k}_3$ axes are random. Micromagnetic cell size $D = 1.5$ nm. There are 32×32 cells in the film (x–y) plane, periodic boundary conditions are used in-plane; © [2011] IEEE. Reprinted, with permission, from Ref. [26]

Table 3.8 Simulation parameters of IrMn/CoFe(PL)/MgO/CoFe(FL) TMR spin valve

Pinned layer	M_s, M_s^a (emu/cm³)	$2K_1$ (erg/cm³)	α_θ	A_1^*, A_2^* (erg/cm)	H_σ (Oe)	$\langle H_{pin} \rangle$ (Oe)
Crystal grain	1,500	5.0×10^4	20	1×10^{-6}	580	600
Grain boundary	95% $\times$ 1,500	0.4×10^4	1	0	580	0
Free layer	M_s, M_s^a (emu/cm³)	$2K_1$ (erg/cm³)	α_θ	A_1^*, A_2^* (erg/cm)	H_σ (Oe)	$\langle H_{pin} \rangle$ (Oe)
Crystal grain	1,400	5.0×10^4	20	1×10^{-6}	150	–
Grain boundary	95 % $\times$ 1,400	0	1	10% $\times 1 \times 10^{-6}$	150	–

The microstructure is simulated and set as the same for the PL and FL, since the TMR multilayer has columnar crystal grains in experiment. The averaged grain size in the simulation is about 11 nm. According to the experiment [25], both the PL and FL have a BCC structure and (100) texture. Thus, the cubic anisotropy is adopted where the anisotropy energy is described by Eq. (2.54). The cubic crystal axes $\hat{k}_1$, $\hat{k}_2$, $\hat{k}_3$ are set in the same direction in a grain but obeying an orientation distribution $f(\theta) = \exp(-\alpha_\theta \sin^2 \theta_k)$ among grains, in which θ_k denotes the angle between the [100] or $\hat{k}_1$ and the z-axis. The projection of the [001] or $\hat{k}_3$ axes of all cells in a plane of the thin film are illustrated in Fig. 3.11b.

The micromagnetic model of TMR spin valve is improved based on the studies of FeCo films [18], in which the magnetic properties in the crystal grain and at the grain boundary are treated separately, as discussed in the last section. The magnetic parameters are listed in Table 3.8. The magnetostriction field H_σ is set along x-axis for cells adjacent to the MgO barrier, both in the PL and the FL. There are several significant different settings in Table 3.8 for TMR spin valve from the parameters in Table 3.7 for a FeCo layer. First, there is a pinning field H_{pin} to describe the bias effect in the PL from the IrMn layer, this term is special for the spin valve thus is not included in Eq. (2.57). The $\langle H_{pin} \rangle$ is an averaged pinning field over all cells adjacent to the antiferromagnetic IrMn layer. Secondly, the exchange at grain boundary is zero in the PL. Thirdly, the anisotropy at grain boundary is zero in the FL. All these three changes of parameters are important to explain the TMR hysteresis.

The PL is directly coupled to an AFM IrMn layer, which results in a horizontal bias field shift H_{EB} for PL in the measured M–H or R–H loop. This exchange bias effect originates from the effects of uncompensated spins in the AFM layer on the Heisenberg exchange at the FM/AFM interface [27], which depends on the relative orientations of the crystal lattices in the FM and AFM grains. The AFM layer is not directly included in the calculation. Instead, the pinning field H_{pin} is introduced only for the micromagnetic cells in the PL adjacent to the AFM layer to depict the exchange bias effect. Moreover, according to the measurement of X-ray magnetic circular dichroism [28], spins at the IrMn/CoFe interface have a probability to be exchange coupled. Hence in our model, only a certain percentage of crystal grains are coupled to the AFM layer. The corresponding pinning field H_{pin} has a wide distribution similar to Eq. (3.7) among the crystal grains that are pinned:

$$p(H_{pin}) = \exp(-\ln(H_{pin}/H_{pin}^0)^2/\beta^2) \exp(-(H_{pin}/H_{pin}^0)^2), \qquad (3.15)$$

with a large distribution coefficient $\beta = 10$. H_{pin}^0 is the average magnitude of H_{pin} among all cells in the crystal grains that are pinned, thus H_{pin}^0 is larger than the parameter $\langle H_{pin} \rangle$ in Table 3.8. Due to the localization and the nonuniformity of H_{pin} magnitude, the H_{pin} in some grains may reach several thousands Oersted, as seen in Fig. 3.12, where H_{pin} is the same for cells in a certain grain but has a distribution among grains. The $\mathbf{H}_{pin}$ is nearly along the x-axis but with an orientation distribution as $\exp(-\alpha_p \sin^2 \phi)$, with $\alpha_p = 5$, and ϕ is the angle between $\mathbf{H}_{pin}$ and x-axis.

The simulated R–H loops of TMR multilayers are given in Fig. 3.13. The scaled resistance change of a TMR spin valve is calculated approximately as:

$$\Delta R/R = \eta \frac{1}{N_x N_y} \sum_{ij} \left[1 - \hat{m}(i, j, 2) \cdot \hat{m}(i, j, 5) \right] \sim \langle 1 - \cos\theta \rangle \qquad (3.16)$$

where $k = 2$ refers to the plane of cells in PL adjacent to MgO and $k = 5$ is for the plane of cells in FL adjacent to MgO barrier; the angle θ represents the angle between the magnetizations in aligned cells of the FL and PL. The R–H loops in Fig. 3.13 are averaged over 20 samples with the same parameters, but different random distributions of H_k and H_{pin} shown in Figs. 3.11 and 3.12 respectively. The simulated total in-plane size of one sample is only $48 \times 48\,\text{nm}^2$ and thus this average is necessary to illustrate the behavior of a large TMR multilayer.

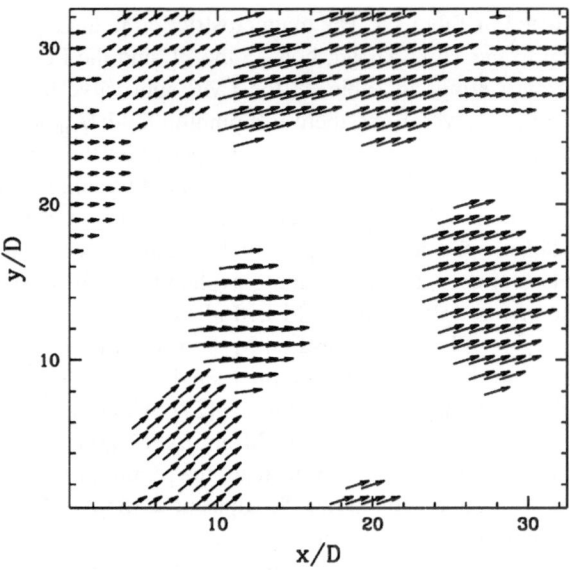

Fig. 3.12 A distribution of H_{pin} shown by arrows in cells, where 50% of grains are pinned; ©
[2011] IEEE. Reprinted, with permission, from Ref. [26]

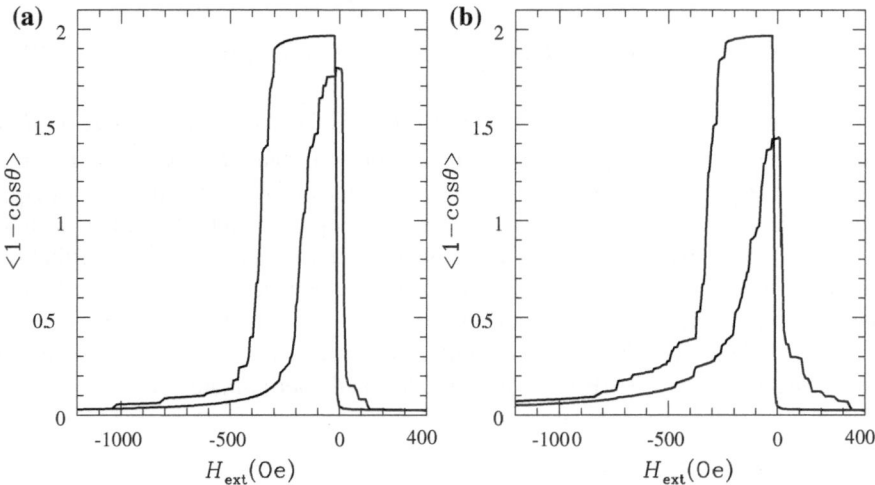

Fig. 3.13 Simulated R–H loops averaged over 20 TMR spin valve samples. **a** 50% grains pinned
in PL by AFM layer; **b** 25% grains pinned in PL by AFM layer; © [2011] IEEE. Reprinted, with
permission, from Ref. [26]

In Figs. 3.13a, b, 50 and 25% of the grains in the PL are coupled by the pinning field
from the IrMn layer, respectively. The measured R–H loop [25] has a long "tail" from
$H_{ext} = -1,000$ Oe to about -500 Oe and a short "tail" from about 0–400 Oe, which

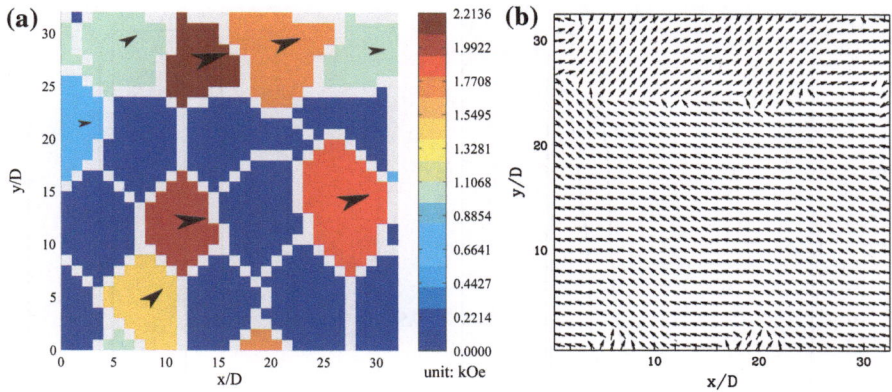

Fig. 3.14 **a** Distribution of H_{pin} by magnitudes, where 50% of grains are pinned; **b** in the curve swept from $H_{ext} = +400$ to -1000 Oe, at $H_{ext} = -800$ Oe, the magnetization pattern in the PL; © [2011] IEEE. Reprinted, with permission, from Ref. [26]

has a shape in between the simulated loops in Figs. 3.13a, b, even closer to the 25% grain-pinning case. Weak inter-grain exchange A_2^* and wide magnitude distribution of H_{pin} in the PL are key factors to simulate this property of the experimental loops. Weak A_2^* in PL (results invariable with $H_{ex}^2 < 100$ Oe, set as 0 in Table 3.8) allows different grains switch relatively independently; and a wide distribution of H_{pin} results in the inclined and biased hysteresis of PL. Comparison between Figs. 3.13a, b shows that with fewer grains biased in PL, the difference of the two curves in the R–H loop is larger, which represents higher local randomness in bias and is unfavorable for the device application.

Analysis of magnetization configuration under different external field shows that nonuniform switching occurs when the external field sweeps. When the sweep of external field starts from 400 Oe and at the very beginning the magnetization of PL and FL is parallel. When the external field H_{ext} is around -15 Oe, the magnetization in the FL reverses, and this results in the increase of $\langle 1 - \cos\theta \rangle$ in the R–H loop. Due to the weak inter-grain exchange coupling A_2^* in the PL, when the external field reduces, nonuniform switching occurs: grains that are not coupled to the AFM, grains with small pinning field and grains with large pinning field switch successively when the external field sweeps from -300 to -1000 Oe. So, the $\Delta R \sim \langle 1 - \cos\theta \rangle$ decreases slowly to nearly zero. In Fig. 3.14b, the magnetization distribution in the PL is given when the external field is -800 Oe. Compared with the corresponding H_{pin} distribution by magnitude in Fig. 3.14a, the magnetization in crystal grains that are not pinned is nearly along $-x$ direction. While the magnetization in grains whose pinning field is relatively large is still at an angle to $-x$ direction or even points to the nearly opposite direction. This plot proves the key roles of weak inter-grain exchange A_2^* and wide distribution of H_{pin} in the PL.

The coercive field of PL and FL and the exchange bias field can be controlled by adjusting parameter of the magneto-elastic field H_σ in PL and FL and the averaged

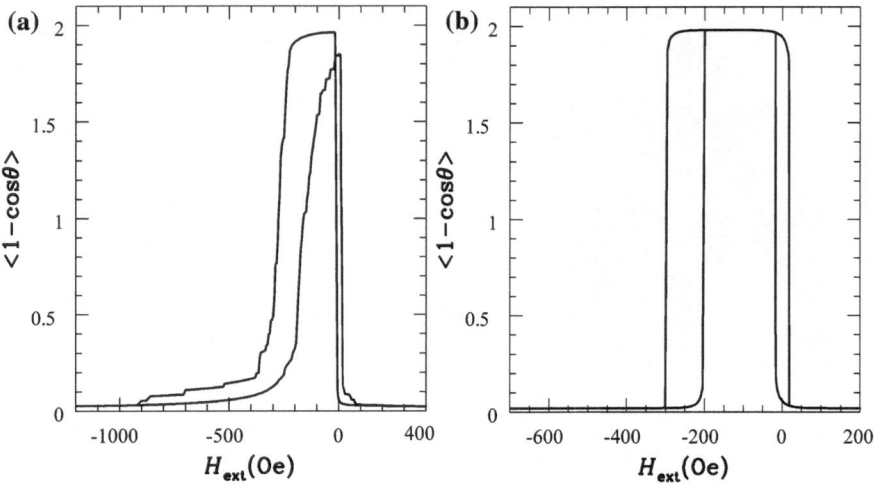

Fig. 3.15 Simulated R–H loops using different parameters ($H_\sigma = 400\,\mathrm{Oe}$ and $\langle H_{\mathrm{pin}}\rangle = 500\,\mathrm{Oe}$ in PL, $H_\sigma = 100\,\mathrm{Oe}$ in FL). **a** 50% of grains pinned in PL; **b** uniform $\mathbf{H}_{\mathrm{pin}}$ applied along x-axis in PL; © [2011] IEEE. Reprinted, with permission, from Ref. [26]

pinning field $\langle H_{\mathrm{pin}}\rangle$ in PL. In Fig. 3.15a, the simulated R–H loops are shown using different parameters of H_σ in FL, H_σ and H_{pin} in PL with 50% of grains pinned; compared to Fig. 3.13a, the coercivity of the FL decreases from 15 to 11 Oe, the coercivity of the PL with respect to the H_{EB} decreases from 96 to 57 Oe, and the bias field H_{EB} itself reduces from 265 to 220 Oe. The behavior of the R–H loop is a result of the interaction between the magnetization in the PL and the FL. The coercive field of the isolated PL is about 63 Oe. Under the interaction with the FL layer, the coercive field of PL reduces to 57 Oe in Fig. 3.13a.

If the pinning field is uniformly applied along x-axis in PL, the simulated R–H loop will be totally different, as seen in Fig. 3.15b, which indicates the significant impact of the wide magnitude distribution of the pinning field H_{pin} at the polycrystalline FM/AFM layer interface. Figure 3.15b agrees with the shape of the measured R–H loop in single crystal TMR [29], although the parameters might be different.

In a summary, the M–H loops along all three directions of magnetic thin films or multilayers can be simulated by micromagnetic models considering the microstructures at different levels. The microstructure reflects the columnar crystal grain distributions in a polycrystalline thin film. The magnetic properties in crystal grains and at amorphous grain boundary are dealt with separately. In a hard magnetic layer, the amorphous phase at grain boundary is usually nonmagnetic; in a soft magnetic layer, the amorphous phase has a slightly smaller saturation and a much lower exchange interaction compared to the crystal phase. In a TMR spin valve, the nonuniformly distributed pinning field and weak inter-grain exchange in the PL is crucial to explain the R–H loop. There are larger-scale defects in microstructures, which might be important for the perpendicular-anisotropy TMR multilayers; for these systems, micromagnetic models with a larger scale are needed.

References

1. Hughes, G.F.: Magnetization reversal in cobalt-phosphorus films. J. Appl. Phys. **54**, 5306–5313 (1983)
2. Victora, R.H.: Micromagnetic predictions for magnetization reversal in CoNi films. J. Appl. Phys. **62**, 4220–4225 (1987)
3. Bertram, H.N., Zhu, J.G.: Micromagnetic studies of thin metallic films. J. Appl. Phys. **63**, 3248–3253 (1988)
4. Feynman, R.P., Leighton, R.B., Sands, M.: The Feynman Lectures on Physics (I). Addison-Wesley, New York (1963)
5. Wang, S.M., Wei, D., Gao, K.Z.: Limits of discretization in computational micromagnetics. IEEE Trans. Magn. **47**(10), 3813–3816 (2011)
6. μMAG—Micromagnetic Modeling Activity Group. NIST Center for Theoretical and Computational Materials Science. http://www.ctcms.nist.gov/rdm/mumag.org.html (2000). Accessed 20 June 2011
7. Zhang, K.M., Hu, X.R., Xie, L., Yuan, J., Zhu, J., Wei, D.: Anisotropy distribution of FePt nanoparticles with twinned structures. IEEE Trans. Magn. **45**(10), 4427–4430 (2009)
8. Bertram, H.N., Zhu J.G.: Fundamental magnetization process in thin film recording media. Solid State Phys. Adv. Res. App. **46**, 271–371 (1992)
9. Wei, D., Liu, B.: Effect of anisotropy magnitude distribution on signal-to-noise ratio. IEEE Trans. Magn. **33**(5), 4381–4384 (1997)
10. Oikawa, T., Nakamura, M., Uwazumi, H., Shimatsu, T., Muraoka, H., Nakamura, Y.: Microstructure and magnetic properties of CoPtCr–SiO_2 perpendicular recording media. IEEE Trans. Magn. **38**(5), 1976–1978 (2002)
11. Piao, K., Li, D.J., Wei, D.: The role of short exchange length in the magnetization processes of L10-ordered FePt perpendicular media. J. Magn. Magn. Mater. **303**, e39–e43 (2006)
12. Ito, H., Shima, T., Takanashi, K., Takahashi, Y.K., Hono, K.: Control of the size for octahedral FePt nanoparticles and their magnetic properties. IEEE Trans. Magn. **41**(10), 3373–3375 (2005)
13. Li, Z.H., Cao, J.W., Wei, F.L., Piao, K., She, S.X., Wei, D.: Micromagnetic analysis of L10 ordered FePt perpendicular recording media prepared by magnetron sputtering. J. Appl. Phys. **102**, 113918 (2007)
14. Cao, J., Cai, J., Liu, Y., Yang, Y., Wei, F., Xia, A., Han, B., Bai, J.: Effect of CrW underlayer on structual and magnetic properties of FePt thin film. J. Appl. Phys. **99**(8), 08F901 (2006)
15. Victora, R.H., Shen, X.: Composite media for perpendicular magnetic recording. IEEE Trans. Magn. **41**(2), 537–542 (2005)
16. Hall, R.C.: Magnetic anisotropy and magnetostriction of ordered and disordered cobalt-iron alloys. J. Appl. Phys. **31**(5), S157–S158 (1960)
17. Okada, Y., Hoshiya, H., Okada, T., Fuyama, M.: Magnetic properties of FeCo multilayered films for single pole heads. IEEE Trans. Magn. **40**(4), 2368–2370 (2004)
18. Wang, S.M., Wei, D., Gao, K.Z.: Initial permeability and dynamic response of FeCo write pole. IEEE Trans. Magn. **46**(6), 1951–1954 (2010)
19. Berkowitz, A.E., Takano, K.: Exchange anisotropy—a review. J. Magn. Magn. Mater. **200**, 552 (1999)
20. Shimatsu, T., Katada, H., Watanabe, I., Muraoka, H., Nakamura, Y.: Effect of lattice strain on soft magnetic properties in FeCo/NiFe(Cr) thin films with 2.4T. IEEE Trans. Magn. **39**(5), 2365–2367 (2003)
21. Bozorth, R.M.: Ferromagnetism. IEEE Press, New York (1993)

22. Jung, H.S., Doyle, W.D., Matsunuma, S.: Influence of underlayer on the soft properties of high magnetization FeCo films. J. Appl. Phys. **93**(10), 6462–6464 (2003)
23. Ruderman, M.A., Kittel, C.: Indirect exchange coupling of nuclear magnetic moments by conduction electrons. Phys. Rev. **96**, 99–102 (1954)
24. Zhang, X.-G., Butler, W.H.: Large magnetoresistance in bcc Co/MgO/Co and FeCo/MgO/FeCo tunnel junctions. Phys. Rev. B **70**, 172407 (2004)
25. Parkin, S.S.P., Kaiser, C., Panchula, A., Rice, P.M., Hughes, B., Samant, M., Yang, S.H.: Giant tunnelling magnetoresistive at room temperature with MgO (100) tunnel barriers. Nat. Mater. **3**, 862–867 (2004)
26. Wang, Y., Wei, D., Gao, K.Z.: Micromagnetic studies on tunneling magnetoresistive spin valves. IEEE Trans. Magn. **47**(10), 2720–2723 (2011)
27. Takano, K., Kodama, R.H., Berkowitz, A.E., Cao, W., Thomas, G.: Interfacial uncompenstated antiferromagnetic spins: role in unidirectional anisotropy in polycrystalline $Ni_{81}Fe_{19}$/CoO bilayers. Phys. Rev. Lett. **79**(6), 1130–1133 (2004)
28. Ohldag, H., Scholl, A., Nolting, F., Arenholz, E., Maat, S., Young, A.T., Carey, M., Stohr, J.: Correlation between exchange bias and pinned interfacial spins. Phys. Rev. Lett. **91**, 017203 (2003)
29. Yuasa, S., Nagahama, T., Fukushima, A., Suzuki, Y., Ando, K.: Giant room-temperature magnetoresistance in single-crystal Fe/MgO/Fe magnetic tunnel junctions. Nat. Mater. **3**(12), 868–871 (2004)

Chapter 4
Domain Structure and Dynamic Process

Abstract This chapter will discuss the micromagnetic simulation of magnetization pattern or domain structures in magnetic devices, both the static domain and dynamic process will be studied. The domain structure is less sensitive to the microstructure compared to the M–H loop; however it is very sensitive to the initial configuration of magnetization and randomness of anisotropy, which strongly reflects the nonlinearity nature of the Landau–Lifshitz equation. The demagnetizing matrices of arbitrary polyhedron cells, as discussed in Chap. 2, will be utilized to study the domain structure in magnetic devices with inclined edges, such as writers in hard disk drives and tips in magnetic force microscope. The limits of cell size in computational micromagnetics will also be discussed.

Keywords Exchange length and micromagnetic cell size · Nonlinearity in Landau–Lifshitz equations · Improved FDM–FFT method · Dynamic switching in write heads · Domain and stray field in Magnetic Force Microscope tip

In Brown's book *Micromagnetics* published in 1963 [1], he summarized the "magnetization curve theory" and "domain theory" before 1960s, which are the main parts of today's micromagnetics. The "magnetization curve theory" or the theory of M–H loop has been discussed for materials in media and heads in Chap. 3; the static and dynamic domain theory will be analyzed in this chapter.

In 1932, Felix Bloch worked out a structure of the boundary wall between two Weiss domains, using an anisotropy energy as the second-order uniaxial anisotropy term in Eq. (2.55) or Eq. (2.56) in a crystal with hexagonal or tetragonal symmetry [2]. Although the common expression of the "Bloch wall", $m_z = \tanh(x/a)$ (the wall width $a = l_{ex}^B = \sqrt{A^*/K_1}$ is just the Bloch exchange length), was given by Landau and Lifshitz, simply using the first term of uniaxial anisotropy [3]. In the traditional "domain theory", the magnetization in a domain is treated as a uniform vector; here we still use the term of "domain theory", however, the space variation of magnetization will be included, down to the size of a micromagnetic cell.

The cross-track width of the main pole tip in writer is scaled below 100 nm in recent years. There are two inclined edges at both side of the main pole. If the FDM–FFT method is used to simulate the writer, zig-zag shapes would appear at the inclined edge, which might influence the dynamic switching property. During the meeting TMRC 2007, Kai-Zhong Gao from Seagate Technology discussed this issue with the author, and we decided to find a way to solve this problem. The solution is to introduce the triangular prism or other polyhedron micromagnetic cells at the edge of an arbitrary-shaped ferromagnetic device [4]. The analytical demagnetizing matrices are derived for these edge cells, as discussed in Sect. 2.2, thus higher accuracy can be achieved with durable computational speed.

The dynamic processes of domain structure, such as switching property and high frequency response, are important in applications of soft magnetic devices such as writers. In the write process, the time scale for one bit is on the order of 1 ns; thus the switching time of a writer should be much faster than 1 ns. Two types of magneto motive force (MMF) are used: one is a simple model using the magnetic pole at the top of main pole tip as MMF, the other is a model of the writer using the current in coil as MMF. The switching properties are similar in these two models, which reveal the importance of the geometrical shape of the main pole tip.

In this chapter, micromagnetic models were built for magnetic devices with arbitrary shape, and comparisons are made for models using just regular mesh or including the microstructures of polycrystalline magnetic thin films . The writer in hard disk drive (HDD) and the magnetic tip in the magnetic force microscope (MFM) would be the main focus of the devices. The domain structure is less sensitive to the microstructure compared to the M–H loop; however it is very sensitive to the initial configuration of magnetization and the randomness of anisotropy, which strongly reflects the nonlinearity nature of the Landau–Lifshitz equation. The improved FDM–FFT micromagnetic methods will be used, where the fast-Fourier-transform (FFT) is still the method to calculate the magneto-static interaction fields among regular cells, which are the majority cells in the main body of a device, the direct sum should be done for demagnetizing fields related to the non-regular edge cells. The effect of the Landau damping constant will also be discussed.

4.1 Reliability and Stability of Domain Calculation

Although micromagnetics has been successful in the engineering applications, there are still difficulties in the accuracy of the micromagnetics theory; the most important problem relates to the way of discretization of micromagnetic cells.

In a micromagnetic model of ferromagnetic metals, the exchange energy or stiffness constant A^* is on the order of 10^{-6} erg/cm between neighbor cells inside a crystalline grain, which is consistent with the exchange interaction constant $J_e = A^* R$ in the Heisenberg model, where R is the distance between neighbor atoms. The main assumption in micromagnetics is that the magnetic moments inside a micromagnetic cell have to rotate uniformly. With two neighbor cells at a distance of $D \sim 1$ nm, the exchange field constant $H_e^0 = 2A^*/(M_s D^2) = 2J_e/(M_s D^2 R)$ is much smaller than the Weiss field $H_E \simeq zJ_e/(M_s R^3) \sim 10^7$ Oe among atoms, where $z \sim 10$ is the number of nearest neighbor atomic spins. Furthermore, with larger cell size D, the difference between H_e^0 and H_E is larger. Therefore there is an upper limit of the discretization cell size D for an accurate domain calculation in micromagnetics.

The FeCo soft magnetic thin films are chosen to study the discretization limits in micromagnetic model [5]. The cuboid device has an in-plane size of 48 × 48 nm and a thickness δ of 6 or 12 nm. The saturation $4\pi M_s$ is 2.4 T, the uniaxial anisotropy constant K_1 is 0 or 50, 000 erg/cm^3, and the anisotropy direction $\hat{k}$ in all cells is supposed to be along one side of the cuboid device. In model A, a regular mesh is utilized and the cubic cell size D is chosen as 2, 3, 4 ($\delta = 12$ nm) and 6 nm respectively. For a fixed D, the exchange constant A^* is altered to determine the critical (smallest) exchange length l_{ex}^0 allowed for an accurate domain calculation.

In Fig. 4.1, when $A^* = 0.96 \times 10^{-6}$ erg/cm (the Néel exchange length $l_{ex}^N = 2.04$ nm), the static domain structures are similar as single-domain with cell sizes of 2 and 3 nm; however, the vortex state starts to appear with a cell size $D = 4$ nm. When $A^* = 1.44 \times 10^{-6}$ erg/cm ($l_{ex}^N = 2.50$ nm), the static domain structures are single-domain with $D = 2$, 3 and 4 nm, as seen in Fig. 4.1; however, the vortex state starts to appear with a cell size of 6 nm (not shown here). In this cuboic device, with typical A^* of ferromagnetic metals, at the center of a vortex, the angle between neighbor moments at a distance D is near 90°, which is an error. Thus, the critical ratio D/l_{ex}^0 for an accurate domain calculation, where the vortex state does not appear, can be found for a chosen cubic cell size D, as shown in Fig. 4.2.

In the work of Rave et al. [6], where 2-D models were studied, it was claimed that the in-plane cell size D_0 should be less than the Néel exchange length l_{ex}^N for an accurate simulation in the soft magnetic materials. In our 3-D micromagnetic model, when the film is very thin ($\delta = 6$ nm), the critical ratio D/l_{ex}^0 is approximately in a range of 2–3 when the D is between 2 and 6 nm; for a thicker film ($\delta = 12$ nm), the value of D/l_{ex}^0 is smaller for a correct domain calculation. The value of D/l_{ex}^0 just changes slightly versus the ratio $Q = K_1/(2\pi M_s)$ in the soft magnetic limit.

It would be interesting to study the reliability of simulation models with polyhedron edge cells included. Two different ways of discretization in the micromagnetic models for the same cuboid device are compared for the static domain calculation and the dynamic switching process. Meanwhile, the reliability of our improved FDM–FFT method for demagnetizing field calculation [4], as introduced in Sect. 2.2, is studied by a comparison between these two discretization models.

In Model A, the cuboid device of 48 × 48 × δ nm^3 is composed of only cubic cells; and in Model B, the same device rotated by 45° is discretized into cubic cells in the main body and polyhedron cells at the edges, which are respectively shown in Fig. 4.3. The cubic cell size D is 3 nm. The uniaxial anisotropy direction $\hat{k}$ in a cell is always supposed to be along one side of the square: along $\hat{e}_x$ for Model A and $\hat{e}_x + \hat{e}_y$ for Model B, and no randomness is involved.

The exchange constant A^* is varied between 0.5 and 2 × 10^{-6} erg/cm. In the exchange field calculation, contributions from all the adjacent cells are considered. In Model A, the exchange field constant H_e is always equal to $H_e^0 = 2A^*/(M_s D^2)$;

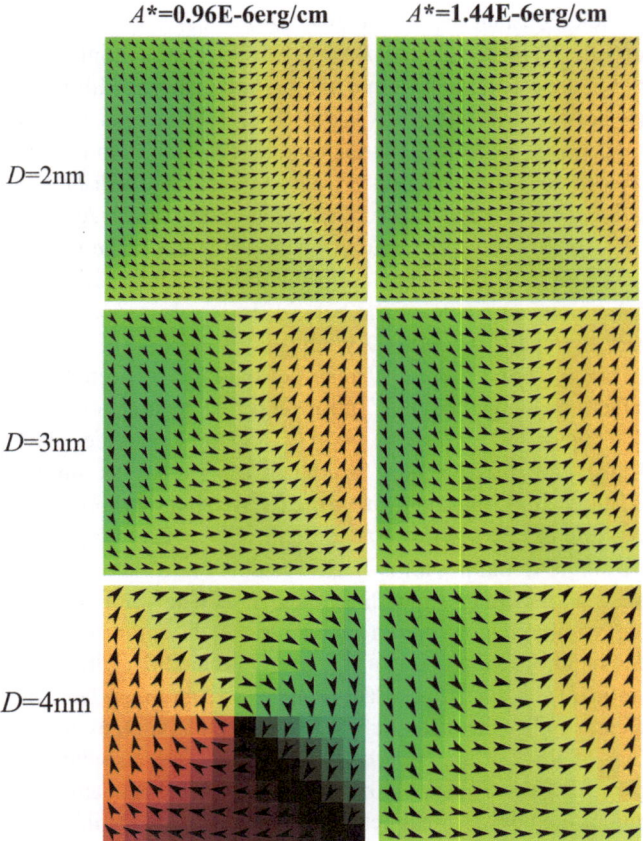

Fig. 4.1 Static domain structures of a cuboid device (thickness $\delta = 12$ nm) with different exchange constant A^* and cell sizes D. The simulated domain pattern may be wrong when D is too large; © [2011] IEEE. Reprinted, with permission, from Ref. [5]

however, as for edge cells in Model B, the constant H_e is proportional to the contact area s of adjacent cells and inversely proportional to the square of the distance $R_{ij} = |\mathbf{r}_i - \mathbf{r}_j|$ between the barycenters of adjacent cells.

In Model B, the total demagnetizing field H_d^B acting on the ith cell, contributed by both the regular cubic cells and prism edge cells, is calculated as:

$$\mathbf{H}_d^B(\mathbf{r}_i) = -4\pi M_s \sum_{j}^{\text{cub}} \tilde{N}_{ij} \cdot \hat{m}_j - 4\pi M_s \sum_{l}^{\text{edge}} \tilde{N}_{il}^{(e)} \cdot \hat{m}_l \tag{4.1}$$

The first term in Eq. (4.1) is the contribution of all regular cells (cubic cells here) and the latter one is that of trapezoid or triangular prism edge cells. The analytical expression of the demagnetizing matrix of cuboid or right-angle triangular prism

Fig. 4.2 In a cuboid device, the scaled critical ratio D/l_{ex}^0 when using different cell size D; © [2011] IEEE. Reprinted, with permission, from Ref. [5]

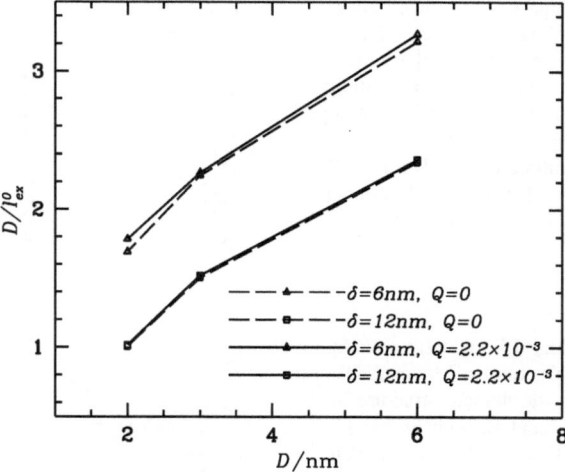

Fig. 4.3 Discretization by different cells in model A (**a**) and model B (**b**) for the same device; © [2011] IEEE. Reprinted, with permission, from Ref. [5]

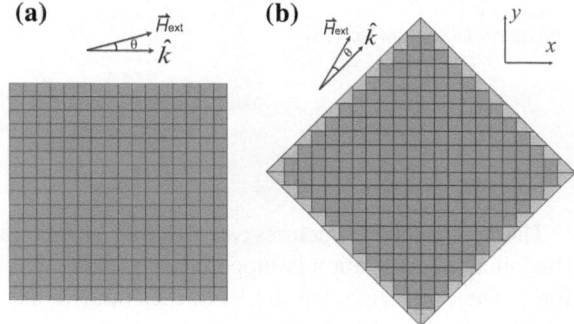

cells have been given Eqs. (2.42) and (2.43) respectively; the demagnetizing matrix of $\tilde{N}^{(e)}$ with trapezoid shape can usually be found by subtraction of the demagnetizing matrices $\tilde{N}^{(t)}$ of two different triangular prism cells. The FFT method is still utilized to calculate the magnetostatic interaction field among cubic cells, and all other terms in Eq. (4.1) are calculated by direct sum for higher accuracy.

The accuracy of the demagnetizing field calculation has been checked in model A. When the magnetization $\hat{m}_i$ is uniform and set along the uniaxial anisotropy direction $\hat{k}$, the calculated demagnetizing fields in each cell have been compared to the analytical demagnetizing field $\mathbf{H}_d^0$ of the whole cuboid device following Table 2.1, which reveals almost no difference. The accuracy of the demagnetizing field calculation is also checked in Model B. If the $\hat{m}_i$ is uniform and set to be along $\hat{k}$, the demagnetizing fields $\mathbf{H}_d^B$ calculated by Eq. (4.1) at the barycenters $\{\mathbf{r}_i\}$ of each cells are compared with the analytical demagnetizing fields $\mathbf{H}_d^0$ of the whole cuboid device situated at corresponding positions $\{\mathbf{r}_i\}$. In Fig. 4.4, the size of black dots displays the component of the demagnetizing fields $\mathbf{H}_d^B$ and $\mathbf{H}_d^0$ in the $-\hat{k}$ direction. The relative differences between them are no more than 10^{-6}.

Fig. 4.4 The distribution of the component of the demagnetizing field along k when the magnetizations are all along the anisotropy field direction: **a** the whole cuboid device: **b** Model B; © [2011] IEEE. Reprinted, with permission, from Ref. [5]

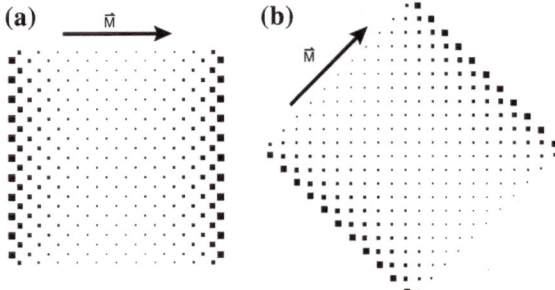

Fig. 4.5 **a** Static domain structure in Model A **b** static domain structure in Model B, with exchange $A^* = 1.44 \times 10^{-6}$ erg/cm, film thickness $\delta = 12$ nm and cubic cell size $D = 3$ nm; © [2011] IEEE. Reprinted, with permission, from Ref. [5]

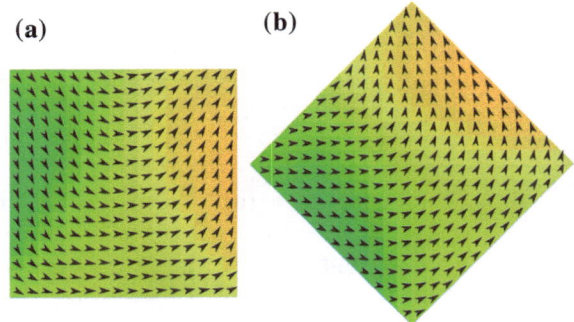

The static domain structures calculated by Model A and B are compared in Fig. 4.5. The initial magnetization is supposed to be parallel to the uniaxial anisotropy direction $\hat{k}$. The static states are similar to each other in the two models, which is actually valid for all studied exchange constant values in the range of 0.5 to 2×10^{-6} erg/cm. The critical ratios $D_{av}/l_{ex}^0 = 2.146$ and 1.462 are also found for Model B with thickness $\delta = 6$ and 12 nm, respectively, where D_{av} is the average cell size and equal to 2.97 nm. These values D_{av}/l_{ex}^0 in Model B are similar to the D/l_{ex}^0 of 2.266 and 1.515 with $\delta = 6$ and 12 nm respectively in Model A, where the cell size D is 3 nm and similar to D_{av} in model B.

To investigate the dynamic switching processes, an external field H_{ext} is applied at an angle $\theta = 10°$ with respect to the uniaxial anisotropy direction $\hat{k}$ in Model A and B respectively, which can be seen in Fig. 4.3. The reversal processes are simulated with the initial magnetization along $-\hat{k}$ in both models. The amplitude value of the external field H_{ext} varies from H_k to $20H_k$, where the crystalline anisotropy field constant H_k equals $2K_1/M_s = 52.36$ Oe.

In Fig. 4.6, the domain structures are shown for $\langle \hat{m} \cdot \hat{k} \rangle = -0.5, 0, 0.5$ in model A and B, respectively. It can be seen that when $\langle \hat{m} \cdot \hat{k} \rangle = -0.5$ or 0, the magnetization distributions are similar in the two models, the respective switching time is also similar. However, when $\langle \hat{m} \cdot \hat{k} \rangle = 0.5$, there are significant difference, basically model B has slower response to the external field than model A.

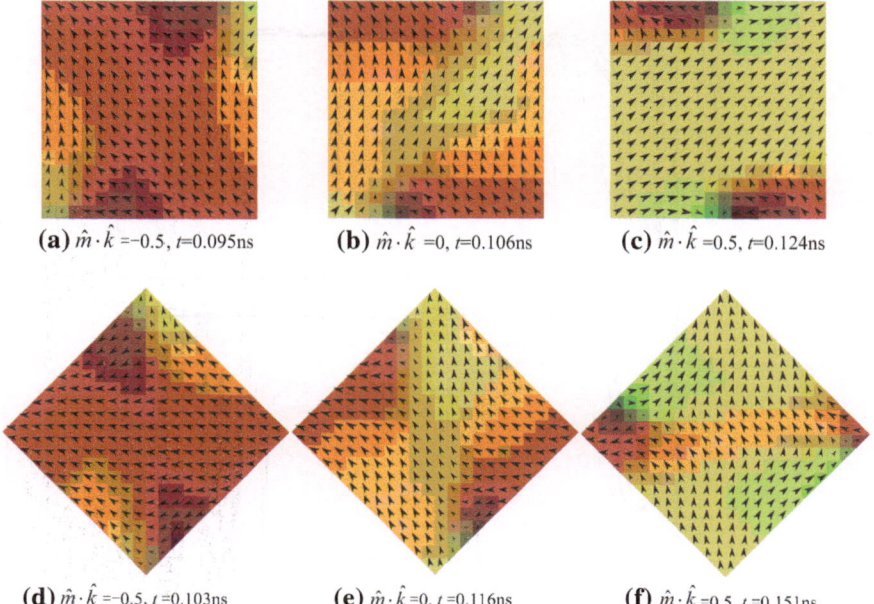

(a) $\hat{m} \cdot \hat{k} = -0.5$, $t = 0.095\,\text{ns}$ **(b)** $\hat{m} \cdot \hat{k} = 0$, $t = 0.106\,\text{ns}$ **(c)** $\hat{m} \cdot \hat{k} = 0.5$, $t = 0.124\,\text{ns}$

(d) $\hat{m} \cdot \hat{k} = -0.5$, $t = 0.103\,\text{ns}$ **(e)** $\hat{m} \cdot \hat{k} = 0$, $t = 0.116\,\text{ns}$ **(f)** $\hat{m} \cdot \hat{k} = 0.5$, $t = 0.151\,\text{ns}$

Fig. 4.6 Domain structures in switching process for Model A (**a–c**) and Model B (**d–f**) with $H_{\text{ext}} = 20H_{\text{k}}$, $\theta = 10°$, $A^* = 1.44 \times 10^{-6}$ erg/cm, thickness $\delta = 12$ nm and cell size $D = 3$ nm

In Fig. 4.7, the average magnetization of the device along $\hat{k}$ is plotted versus time with a fixed magnitude $H_{\text{ext}} = 20H_{\text{k}}$ and different angle $\theta = 0°, 5°, 10°, 15°, 20°$ of the external field with respect to $\hat{k}$. Most of switching properties are similar; however, when $\theta = 0°$, there is a large difference between model A and B. There is a small error (about 0.17%) due to the usage of the polyhedron edge cells in model B, and this error can have significant influence when the external field is parallel/antiparallel to the uniaxial anisotropy, which is related to the $emm \times \mathbf{H}_{\text{eff}}$ form and the nonlinearity of the Landau–Lifshitz equation.

In Fig. 4.8, the average magnetization of the device along $\hat{k}$ is plotted versus time with a fixed angle $\theta = 10°$ and different magnitude $H_{\text{ext}}/H_{\text{k}} = 1, 5, 10, 15, 20$ of the external field. It can be observed that the reversal processes are similar for both models, especially with larger external fields. Actually, even with larger cells, if the external field is large enough ($15H_{\text{k}}$ in our model), despite different static domain structures of the two models, the switching properties can still be analogous.

The results in this section test and verify the correctness of the improved FDM–FFT method, which includes the irregular micromagnetic cells at the edges of a device. In a model, we should use regular cells as much as possible, and the irregular polyhedron cells can be used at the inclined surfaces.

The limit of the cell size is also found if we just use a regular mesh. It should be emphasized that, the limits of discretization can vary in different devices. We have already seen the effect of film thickness in Fig. 4.2. Actually, in devices with inclined

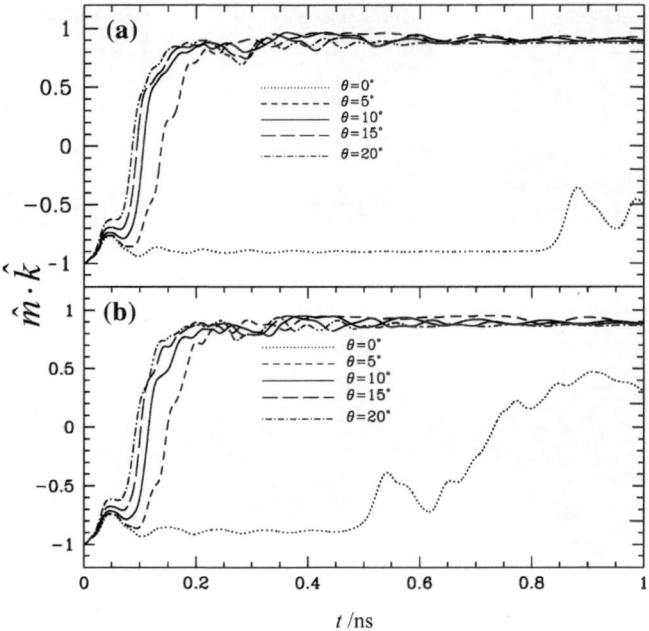

Fig. 4.7 Reversal processes with different angle θ in Model A (**a**) and Model B (**b**) $H_{\text{ext}} = 20H_{\text{k}}$, $A^* = 1.44 \times 10^{-6}$ erg/cm, thickness $\delta = 12$ nm and cubic cell size $D = 3$ nm

Fig. 4.8 Reversal processes with different amplitudes of external fields applied at an angle $\theta = 10°$ in Model A (**a**) and Model B (**b**) with $A^* = 1.44 \times 10^{-6}$ erg/cm and cubic cell size $D = 3$ nm; © [2011] IEEE. Reprinted, with permission, from Ref. [5]

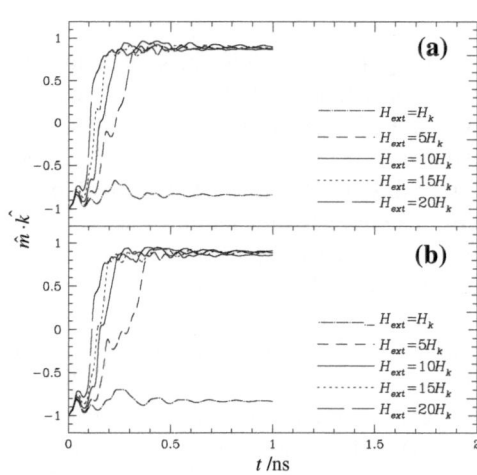

edges, such as main poles in hard disk writers, the critical ratio D/l_{ex}^0 can be even larger and we will see this effect in the next section.

Table 4.1 Simulation parameters in muMAG standard Problem No.3 (assume $M_s = 798$ emu/cc)

L (nm)	D (nm)	N_x, N_y, N_z	α_θ	dt (s)	Error max$\{|d\hat{m}|\}$	A^* (erg/cm)	l_{ex}^N (nm)
30–90	L/N_x	8, 16	$\infty, 10, 0$	10^{-13}	10^{-6}	1×10^{-6}	5

4.1.1 muMAG Standard Problems No. 3

In Sect. 3.1.2, the muMAG standard problem No.2 has been discussed to verify the correctness of demagnetizing field calculation. In this subsection, the muMAG standard problem No.3 will be analyzed, which was proposed by Alex Hubert, University of Erlangen–Nuremberg.

Standard problem No.3 is to calculate the single domain limit of a cubic magnetic particle [7]. The uniaxial anisotropy energy K_1 is set to be $0.1 K_m$ with the easy axis parallel to a principal axis of the cube (say z-axis), and $K_m = 2\pi M_s^2$ is the magnetostatic energy density related to the Néel exchange length $l_{ex}^N = \sqrt{A^*/K_m}$. The limit of the cube size L_0 is to be found for a single-domain "flower" state and a multi-domain "vortex" state, the respective energy terms are also required (Table 4.1).

The domain structure depends on the initial condition, as seen in Fig. 4.9. The anisotropy field $H_k = 2K_1/M_s = 1002$ Oe is set along the z-axis, there is no randomness in the system. The body center of the cube is chosen as the origin. If the initial magnetization is set as $m_z = +1$ for $x < 0$ and $m_z = -1$ for $x > 0$, the static domain structure is always a vortex state, with its vortex center line pointing to the y-axis. If the initial magnetization is set as $m_z = 1$, and the grain size L is less than the critical size L_0, the static domain structure is a near-single-domain flower state, where along a fixed line (x_0, y_0) all the magnetization projections in the $x - y$ plane should coincide in the same direction and parallel to the $x_0\hat{e}_x + y_0\hat{e}_y$ direction.

The critical size of single to multi-domain transition is $L_0 = 41.88$ nm$= 8.38 l_{ex}^N$. When $L = 41.9$ nm, the scaled energy density $E_v/K_m = 0.304073$ of the vortex state is slightly lower than the $E_f/K_m = 0.304136$ of the flower state, where the exchange energy is largely increased but the magnetostatic interaction energy is decreased a little further in E_v with a vortex structure. When the grain size $L > L_0$, the most stable state is the vortex state; however, there are also two meta-stable multi-domain states with the initial condition $m_z = 1$. One is called the twisted flower state I, where along a fixed line (x_0, y_0) the magnetization projections in the $x - y$ plane have a twist; the other is called the twisted flower state II, where along a fixed line (x_0, y_0) all the magnetization projections in the $x - y$ plane are parallel, and are like rotational lines.

Actually the limit of single-domain state was an old question in micromagnetics. Brown used to calculate the single-domain limit for the tape recording magnetic particle γ-Fe_2O_3, and his conclusion was that there exist a size region where both the single-domain state and the multi-domain state exist.

Thus we would like to extend the muMAG standard problem No.3 and include the randomness in calculation, which is more practical. First, if we just introduce a slight randomness in the initial condition of magnetization ($m_z \simeq 1$), as labeled by

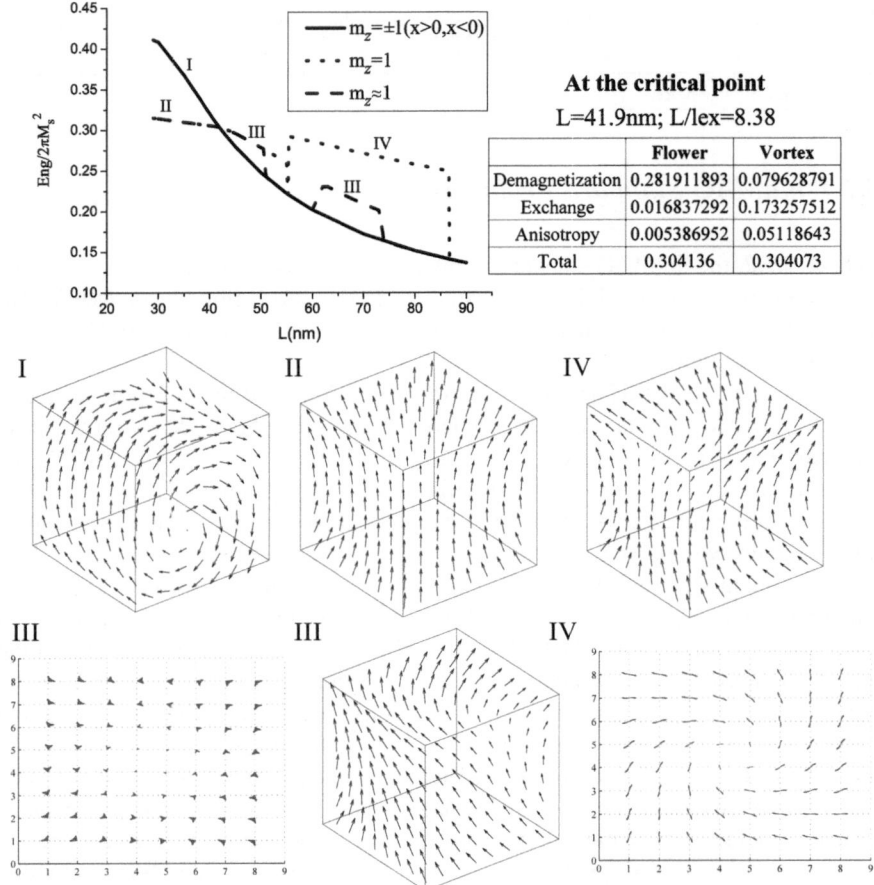

Fig. 4.9 Scaled energy versus grain size L, with the related domain structures: I vortex state; II near-single-domain flower state; III twisted flower state I; IV twisted flower state II

the dashed line in the energy plot of Fig. 4.9, the twisted flower state II will disappear; furthermore, when $L > 55$ nm, the grain will stay in the vortex state or the twisted flower state I with a random probability if a random seed is chosen.

A more realistic situation is to include the orientation distribution of anisotropy fields $\mathbf{H_k}$ in different micromagnetic cells. If the 3-D orientation distribution in Eq. (3.4) is chosen, $\{\mathbf{H_k^i}\}$ distribute isotropically when the distribution coefficient $\alpha_\theta = 0$, and $\{\mathbf{H_k^i}\}$ are well-oriented along z-axis with a large coefficient such as $\alpha_\theta = 10$. The initial magnetization distributes randomly in all cells in the grain. The simulation result is shown in Fig. 4.10. When the grain size $L \leq 36$ nm, the static domain state is always the single-domain flower state. When the grain size 36 nm $< L < 44$ nm, both the single-domain flower state and the multi-domain vortex state can appear, but the probability of the single-domain state decreases with

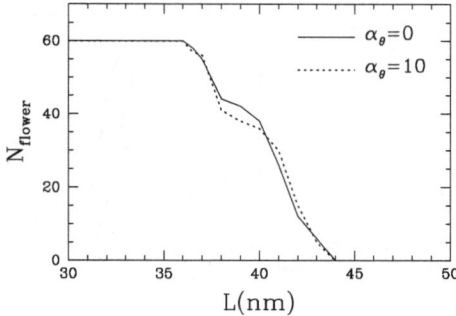

Fig. 4.10 Number of flower states in 60 tries of random H_k distributions, with random initial magnetization in cells: $\alpha_\theta = 0$ for isotropic and $\alpha_\theta = 10$ for well-oriented H_k distribution

Table 4.2 Simulation parameters in muMAG standard Problem No.4 (anisotropy $K = 0$)

D_x, D_y (nm)	D_z (nm)	α	γ (Oe^{-1}s^{-1})	dt (s)	Error max$\{|d\hat{m}|\}$	A^* (erg/cm)	M_s (emu/cc)
2.5, 5	3	0.02	1.75866×10^7	10^{-13}	10^{-6}	1.3×10^{-6}	800

larger size L. Therefore in the size region $L = 7.2 - 8.8l_{ex}^N$, the single-domain state and the multi-domain state co-exist in the cube grain, and this conclusion does not depend on the orientation distribution of anisotropy field. This is a similar conclusion with Brown's study of the single-domain limit for γ-Fe$_2$O$_3$ particles.

4.1.2 muMAG Standard Problems No. 4

muMAG standard problem No.4 focused on the dynamic aspects of micromagnetic computations in a $500 \times 125 \times 3$ nm^3 cuboid device. This problem was brainstormed by Robert McMichael, Roger Koch and Thomas Schrefl, and proposed by Jason Eicke and Robert McMichael [7]. The initial state is an equilibrium s-state, obtained after applying and slowly reducing a saturating field along the [1,1,1] direction to zero (Table 4.2). The reversal processes are run for two different applied fields: 250 Oe at 170° w.r.t. x-axis and 360 Oe at 190° w.r.t. x-axis, and the magnetization components along x,y,z-axes are plotted versus time in Fig. 4.11 respectively.

4.2 Static Domain and Reversal Property of Write Head

SPT head is the main type of design for perpendicular recording write head, which was brought up by Shun-ichi Iwasaki and Yoshihisa Nakamura early in 1977 [8]. To write information in the medium at high speed, the write head itself is also required to resume good reversal properties. In this section, the improved FDM–FFT method will be used by including edge cells in the micromagnetic models of writers [4].

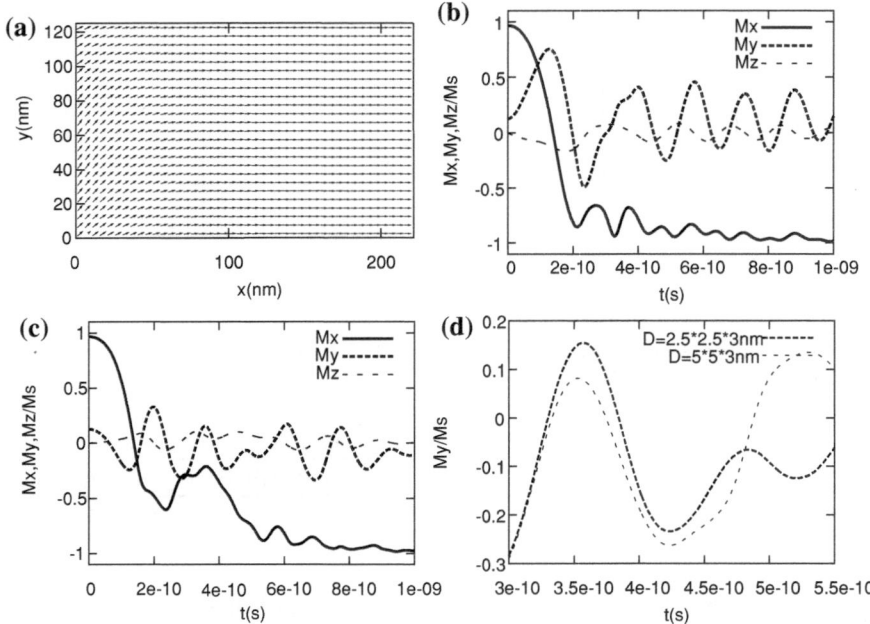

Fig. 4.11 Static domain and reversal property of a $500 \times 125 \times 3 \, \text{nm}^3$ cuboid device **a** part of static domain ($0 < x < 200 \, \text{nm}$); **b** reversal property with $H_{\text{ext}} = 250 \, \text{Oe}$, applied at an $170°$ w.r.t. x-axis; **c** reversal property with $H_{\text{ext}} = 360 \, \text{Oe}$, applied at an $190°$ w.r.t. x-axis; **d** same as **c**, but with a closer look at the controversial part, when two in-plane sizes $D_x = 2.5, 5 \, \text{nm}$ are used

A great amount of research has been done to study the effects of SPT geometry parameters, or magnetic parameters on dynamic processes of write head, where the FDM–FFT micromagnetic models with regular mesh was utilized. However, when the main pole tip of SPT head is scaled down to nanometer region, the inclined edge in the main pole tip will have a sawtooth shape when modeled by the regular mesh. It would be important to clarify the effect of the edge by using the improved FDM–FFT method introduced in Sect. 2.2.

In Fig. 4.12, the reversal properties of the main pole tip in SPT head are compared by two models: "sawtooth edge" model where all cells are cubic and "smooth edge" model in which there are cubic cells and triangular prism edge cells [4].

The simulation parameters of main pole tip in Fig. 4.12 are listed in Table 4.3. In this case the micromagnetic cell size $D = 10 \, \text{nm}$ is relatively large, thus the anisotropy is set as an effective uniaxial anisotropy along the cross-track z-axis with a very small anisotropy field constant $H_k = 2K_1/M_s = 5.23 \, \text{Oe}$. We will compare this case with the main pole tip model including the microstructure later in this section. A MMF or driving magnetic field of $B_s = 4\pi M_s$ is applied on the top surface of the main pole tip to study the reversal properties. The magnetic pole is controlled by the coil thus it rises as $1 - \exp(-t/\tau)$, and here the rising time $\tau = 40 \, \text{ps}$ is chosen for the study.

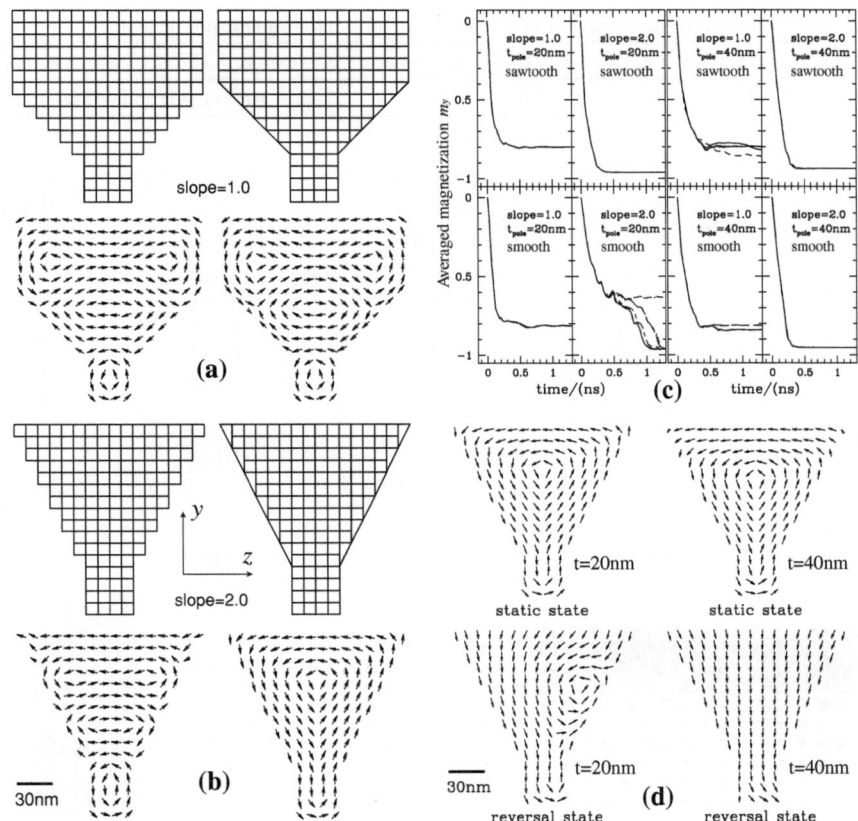

Fig. 4.12 Static domain and reversal property of main pole tip calculated by model A with sawtooth edge (all cubic cells) and model B with smooth edge (including triangular edge cells) **a** static domain with edge slope 1.0, $t_{pole} = 20$ nm; **b** static domain with edge slope 2.0, $t_{pole} = 20$ nm; **c** reversal properties in main poles with thickness $t_{pole} = 20$ and 40 nm, 7 random tries of $\mathbf{H_k}$; **d** comparison of static and reversed domain with edge slope 2.0, $t_{pole} = 20, 40$ nm

The pole tips with an edge slope of 1.0 and 2.0 are simulated to evaluate the effects of sawtooth shape on reversal properties. Reversal properties of model A (sawtooth-edge) and model B (smooth-edge) are similar, as seen in Fig. 4.12c, except when the film thickness $t_{pole} = 20$ nm, the switching process in model B (with both cubic and triangular prism cells) is slower with edge slope 2.0, actually this delay in reversal process is true for thin main pole with slope larger than 1.0.

The difference in the static domain structure can explain this delay in reverse. The inclusion of triangular prism cells will result in different static state in model B for the main pole with an edge slope larger than 1.0, as shown in Fig. 4.12a, b. Furthermore, when the film thickness $t_{pole} = 20$ nm is smaller than the main pole tip width $W = 40$ nm, the demagnetizing field will force the magnetization to stay in film plane, and the magnetostatic interaction of the narrow tip throat will cause

Table 4.3 Simulation parameters of FeCo main pole tip using magnetic poles as MMF

D (nm)	t_{pole} (nm)	M_s (emu/cc)	K_1 (erg/cc)	α_θ (along z)	A^* (erg/cm)	H_e (Oe)	MMF: B_s (Oe)
10	20, 40	1910	5,000	8	0.95×10^{-6}	1,000	24,000

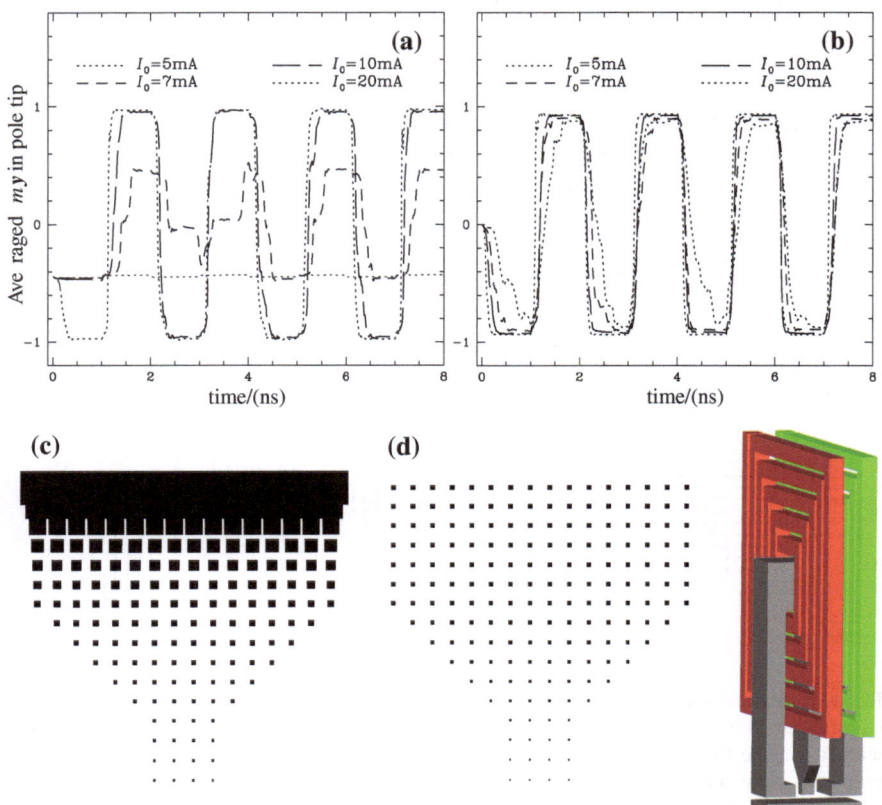

Fig. 4.13 Reversal property using model B in the whole write head driven by current $I_s = 5, 7, 10, 20$ mA in the two coils with pole tip thickness $t_{pole} = 20$ nm **a** and $t_{pole} = 40$ nm **b**; comparison of the MMF field component H_y driven by the magnetic pole **c** or the current in coil **d**

vortex in the reversal process of main pole tip, as shown in Fig. 4.12d, which is hard to be erased and causes the delay in head field reversal.

The switching property of the whole writer, which are driven by current I_s in two coils creating opposite magnetic field, can also be simulated [9]. The field strength H_y perpendicular to the medium is one order higher when using the magnetic pole as MMF than using the current as MMF, as compared in Fig. 4.13c and d; however, the switching time of the main pole with $t_{pole} = 40$ nm is in the range of 1.0–0.1 ns when using current $I_s = 5 - 20$ mA, respectively, as shown in Fig. 4.13b, still on the same order as the reversal time in Fig. 4.12c.

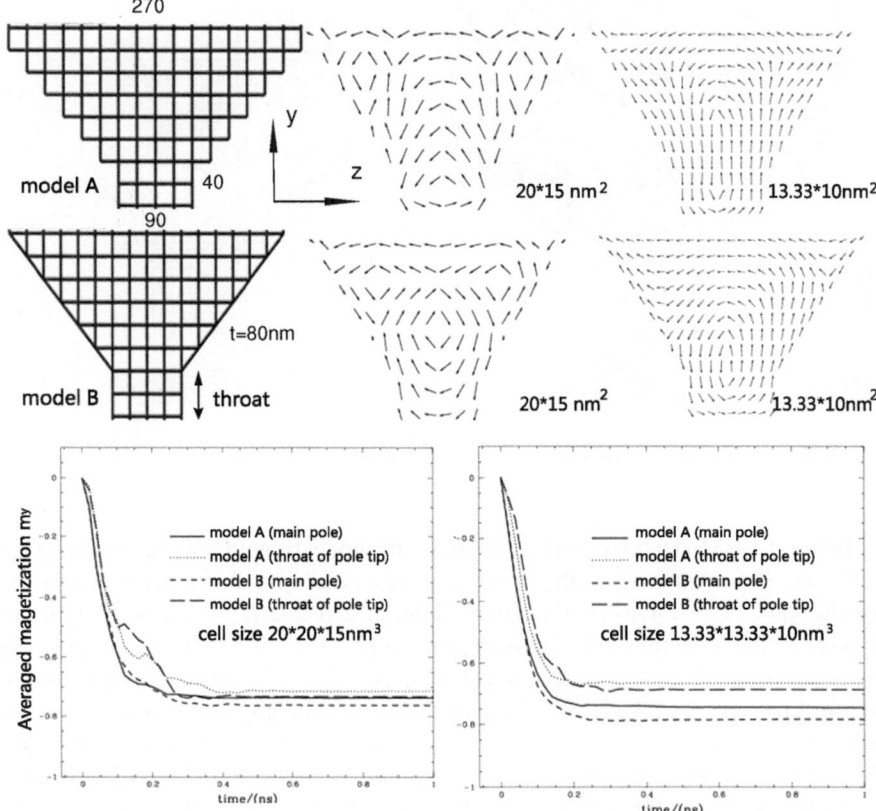

Fig. 4.14 Static domain and reversal property of main pole tip calculated by model A and model B with edge slope 4/3, t_{pole} = 80 nm **a** static domain found by model A with different cell sizes; **b** static domain found by model B with different cell sizes; **c** reversal properties with cell size $D_x \times D_y \times D_z = 20 \times 20 \times 15$ nm^3; **d** reversal properties with cell size $\frac{40}{3} \times \frac{40}{3} \times 10$ nm^3

The Bloch exchange length is $l_{ex}^B = \sqrt{A^*/K_1} \simeq 138$ nm, when the pole tip width W and main pole film thickness t_{pole} is much less than l_{ex}^B, the reversal process will be delayed due to the strong demagnetizing field at the pole tip. A low throat will help the switching. With a very large current $I_s = 20$ mA, the switching time is very short, but the undershoot of head field is very large, which is not acceptable for high density recording; thus the suitable switching current is $I_s = 7, 10$ mA.

The shape of the main pole tip has to be optimized to obtain minimum switching time, because the "quota" time for one bit is only around 1ns for recording in 1 Tb/in^2 hard disks. The switching time of the main pole tip with $h_{throat} = t_{pole} = 40$ nm is 0.4–0.5 ns, as seen in Fig. 4.12c, which is too long. One of the key points in design is to decrease the throat height h_{throat}, say down to less than 1/2 of the cross-track width $W = 90$ nm; the other point is to use a main pole with a long inclined edge with a proper slope, such as 4/3, as shown in Fig. 4.14.

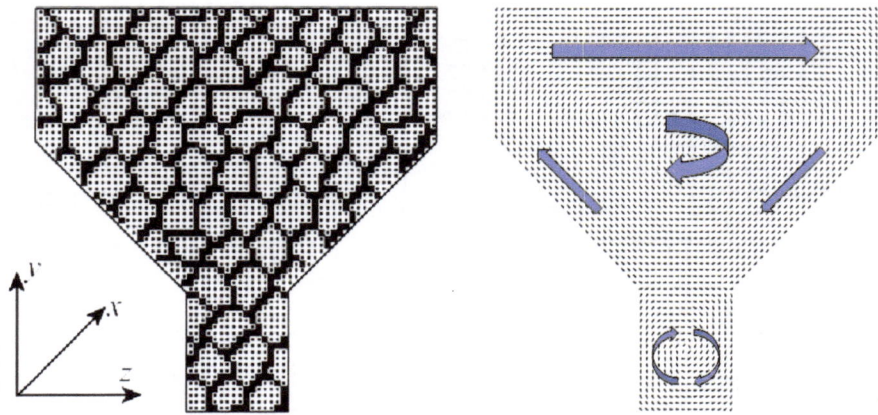

Fig. 4.15 Static domain structure in a main pole tip calculated with microstructure included

In Fig. 4.14, Model A (with all cubic cells) and model B (with triangular prism edge cells) are also compared for the optimized main pole tip. All magnetic parameters are the same as in Table 4.3. The micromagnetic cell size $D_x \times D_y \times D_z$ is set as $D \times D \times \frac{4}{3}D$ to simulate the inclined edge with a slope of 4/3. The pole tip width $W = 90$ nm can be larger than the typical track width $W_t \simeq 20$ nm in a hard disk above 1Tb/in^2, because the Shingled Writing Recording technology (a newly written track partially overlaps the previous track) is applied in current HDD using continuous thin film media. The reversal properties are similar in model A and model B, and the switching is even faster in model B with a smooth edge, where the switching time ($\langle m_y \rangle$ reach 90% of the ultimate average magnetization m_y^0) is only 0.12 ns, much faster than those in Fig. 4.12. Also for all pole tip thickness $t_{pole} = 80-160$ nm, this 0.12 ns switching time of $\langle m_y \rangle$ or head field is stable in model B.

The reliability of domain simulation with different size is an interesting topic. As analyzed in Sect. 4.1, strictly speaking, the cell size D can not be too large compared to the Néel exchange length l_{ex}^N for accurate domain calculation. The $l_{ex}^N = \sqrt{A^*/(2\pi M_s^2)}$ is about 3 nm, thus the $D = 10, 13.3, 20$ nm used in Figs. 4.12 and 4.14 may be too large. For example, when $D = 20$ nm there are extra "minor" vortices appear, which are not accurate compared to the domain structure with $D = 13.3$ nm, as seen in Fig. 4.14b. Actually the simulated domain structure is stable and similar to the domain with $D = 13.3$ nm in Fig. 4.14 when smaller cell size is used. The domain structure simulated with $D = 10$ nm in Fig. 4.12a is also quite close to the accurate domain structure calculated with the microstructure included ($D = 2$ nm), as shown in Fig. 4.15; thus, the domain structure does not rely on the microstructure as sensitive as the M–H loop [10]. The reason is that the strong demagnetizing field caused by two inclined edges and the throat of the pole tip is deterministic for the domain structure in a device like the main pole tip.

4.3 Domain and Field of Magnetic Force Microscope Tip

The MFM, invented by Martin and Wickramasinghe in 1987 [11], is a powerful tool to analyze the fine magnetic domain structures of magnetic materials, especially the magnetic recording media and heads in the recording industry. The resolution of the MFM image, which was in the submicron region in the early days, has been improved to the nanometer scale in recent years. Standard MFM techniques only give qualitative pictures of the magnetic field distribution derivatives. The intrinsic difficulties exist in interpreting the MFM images and relating these "raw" images to the quantitative information of the underlying magnetization patterns and magnetic field distributions.

The point probe model brought up by Hartmann [12] is a simple model often used in the MFM experiments. This model is helpful in estimating the magnetic field of a MFM tip and is also valuable in separating the magnetic contribution from other tip–sample interactions such as van der Waals forces. However, since the magnetization pattern in tip is unknown, the effective magnetic dipole and its exact position are not well discussed in the point probe model.

In this section, the domain structure and the related stray field of CoCr or FeCo MFM tips will be analyzed by the improved FDM–FFT micromagnetic method, where two or more types of polyhedron edge cells will be introduced to simulated the inclined edges of a tip, with a shape of pyramid or pyramid with a hole. The effective dipole and its position are evaluated for different tips.

The position of the effective dipole in a MFM tip can be determined by the multi-pole expansion theory in classical electromagnetism. The magnetic moment distribution $\mathbf{M}(\mathbf{r}_i) = M_s \hat{m}_i$ in a MFM tip can be calculated by the micromagnetic model. In a distance, the magnetic field of an arbitrary magnetization distribution $\mathbf{M}(\mathbf{r})$ can be viewed as a sum of contributions from the effective dipole μ, quadrupole $\tilde{D}$ and other higher order terms. It is reasonable to determine the position of effective dipole by the rule that quadrupole $\tilde{D}$ should be zero, when the origin of the coordinate system is set at the position of the dipole; then, far from the MFM tip, the multi-pole expansion of the magnetic field is closest to the dipole term.

In a MFM tip as illustrated in Fig. 4.16, there is an effective magnetic charge distribution $\rho_m(\mathbf{r}) = -\nabla \mathbf{M}(\mathbf{r})$. The multipole expansion of the magnetostatic potential Ψ_m of a MFM tip at a point $\mathbf{r} = (x, y, z)$ can be written as

$$\Psi_m(\mathbf{r}) = \int d^3\mathbf{r}' \frac{\rho_m(\mathbf{r}')}{|\mathbf{r} - \mathbf{r}'|} = \Psi_m^{(0)}(\mathbf{r}) + \Psi_m^{(2)}(\mathbf{r}) + \Psi_m^{(4)}(\mathbf{r})... \qquad (4.2)$$

where the first term $\Psi_m^{(0)}(\mathbf{r}) = \left(\int d^3\mathbf{r}' \rho_m(\mathbf{r}') \right) / r$ must be zero since there is no intrinsic magnetic charge. The second term is the magnetic dipole term, which is the non-vanishing first order approximation of the potential Ψ_m:

$$\Psi_m^{(2)}(\mathbf{r}) = \int d^3\mathbf{r}' \rho_m(\mathbf{r}') \frac{\mathbf{r} \cdot \mathbf{r}'}{r^3} = \frac{\mathbf{r} \cdot \mu}{r^3} \qquad (4.3)$$

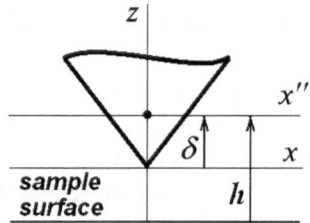

Fig. 4.16 Point probe model in a MFM tip. The sample surface is parallel to the $x - y$ plane; the effective magnetic pole is at the origin of the (x'', y'', z'') axes, where y'' is parallel to y axis and z' coincides with z axis; © [2010] IEEE. Reprinted, with permission, from Ref. [13]

The magnetic dipole $\mu = \int d^3 r' \rho_m(\mathbf{r}') \, \mathbf{r}' = \int d^3 \mathbf{r}' \mathbf{M}(\mathbf{r}')$ is independent of the choice of the coordinate system in Fig. 4.16.

The third term in Eq. (4.2) is the magnetic quadrupole term:

$$\Psi_m^{(4)}(\mathbf{r}) = \int d^3 \mathbf{r}' \rho_m(\mathbf{r}') \left[-\frac{r'^2}{2r^3} + \frac{3}{2} \frac{(\mathbf{r} \cdot \mathbf{r}')^2}{r^5} \right] = \frac{\mathbf{r} \cdot \tilde{D} \cdot \mathbf{r}}{2r^5} \qquad (4.4)$$

The magnetic quadrupole $\tilde{D} = \int d^3 \mathbf{r}' \rho_m(\mathbf{r}') \, (3\mathbf{r}'\mathbf{r}' - r'^2 \mathbb{1})$. Note that in Fig. 4.16, the integrating vector $\mathbf{r}' = \mathbf{r}'' + \delta$ is defined in the coordinate system (x, y, z). In the coordinates $\mathbf{r}'' = (x'', y'', z'')$ with the effective dipole located at the origin, the quadrupole matrix element $\tilde{D}_{33}^{(new)}$ should be zero; then the position δ of the dipole can be found by the formula $\delta_z = D_{33}/(4\mu_z)$:

$$\tilde{D} = \int d^3 \mathbf{r}'' \rho_m(\mathbf{r}'') \left(3(\mathbf{r}'' + \delta)(\mathbf{r}'' + \delta) - (\mathbf{r}'' + \delta)^2 \mathbb{1} \right)$$

$$= \tilde{D}^{(new)} + 3\mu\delta + 3\delta\mu - 2(\mu \cdot \delta)\mathbb{1} \qquad (4.5)$$

$$D_{33} = \int d^3 \mathbf{r}' \left(6M_z(\mathbf{r}')z' - 2\mathbf{M}(\mathbf{r}') \cdot \mathbf{r}' \right) \simeq 4\mu_z \delta_z \qquad (4.6)$$

where it is reasonable to assume that the total magnetic dipole μ of a MFM tip is approximately in the z-direction. Since only the $(3, 3)$ element of the $\tilde{D}^{(new)}$ quadrupole matrix is required to be zero, far from the MFM tip, the true tip field calculated by the micromagnetic model may not be totally identical to the dipole term, which can be verified from the data provided in the following sections.

A micromagnetic model is developed, with two geometrical shapes of magnetic materials in a tip, pyramid or pyramid-with-a-hole (a tip with coating), as illustrated in Fig. 4.17 respectively. The pyramid-with-a-hole tip is closer to the geometry of a true MFM tip. Three types of micromagnetic cells are considered in the model, which include rectangular cells inside the tip body, triangular prism cells on the four side surfaces and quadrangle pyramid cells at the four edges, as labeled in the pyramid part of Fig. 4.17. The magnetostatic interaction contributed by both the triangular prism

Fig. 4.17 Micromagnetic models **a** micromagnetic cells in a pyramid-shaped tip (model A); **b** MFM tip made by magnetic coating (model B) and micromagnetic cells in the cross-section; © [2010] IEEE. Reprinted, with permission, from Ref. [13]

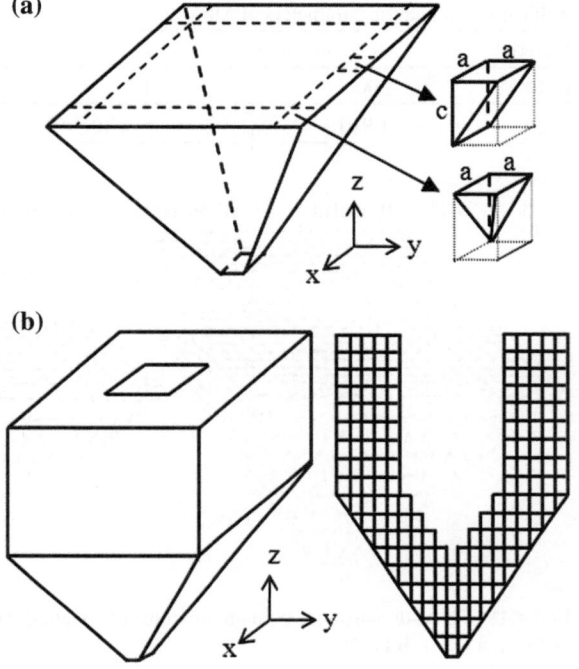

cells and the quadrangle pyramid cells has to be calculated by a direct summation. The total magnetostatic interaction field is:

$$\mathbf{H}_{\mathrm{m}}(\mathbf{r}_i) = -B_{\mathrm{s}} \sum_j^{\mathrm{cub}} \tilde{N}_{ij} \cdot \hat{m}_j - B_{\mathrm{s}} \sum_k^{\mathrm{tri}} \tilde{N}_{ik}^{(\mathrm{t})} \cdot \hat{m}_k - B_{\mathrm{s}} \sum_l^{\mathrm{quad}} \tilde{N}_{il}^{(\mathrm{q})} \cdot \hat{m}_l \qquad (4.7)$$

The demagnetizing matrix contributed by a cuboid cell or a triangular prism cell has been given in Eqs. (2.42) and (2.43) respectively. The demagnetizing matrix of a quadrangle pyramid cell can be calculated in a similar way:

$$\tilde{N}_{ik}^{(\mathrm{q})} = \sum_{l=1}^{1} \tilde{R}_l \cdot \tilde{N}_{\mathrm{s}}^{\mathrm{rec}} \left(\tilde{R}_l^T \cdot (\mathbf{r}_i - \mathbf{r}_k - \delta_l) \right) \cdot \tilde{R}_l^T \qquad (4.8)$$

$$+ \sum_{n=1}^{4} \tilde{R}_n^{(\mathrm{q})} \cdot \tilde{N}_{\mathrm{s}}^{\mathrm{tri}} \left(\tilde{R}_n^{(\mathrm{q})T} \cdot (\mathbf{r}_i - \mathbf{r}_k - \delta_n) \right) \cdot \tilde{R}_n^{(\mathrm{q})T} \qquad (4.9)$$

where $\tilde{N}_{\mathrm{s}}^{\mathrm{rec}}$ or $\tilde{N}_{\mathrm{s}}^{\mathrm{tri}}$ is the demagnetizing matrix contributed by magnetic poles on a rectangular or a right-angle triangular surface located in a fixed plane (say $z' = z_0$), as given in Tables 2.2 and 2.3 respectively.

Table 4.4 Simulation parameters of MFM tips

Material	M_s (emu/cm^3)	K_1 (erg/cm^3)	A^* (erg/cm)	α_θ
CoCr	300	1×10^6	2.25×10^{-7}	0
FeCo	1,910	5×10^3	1.44×10^{-6}	0

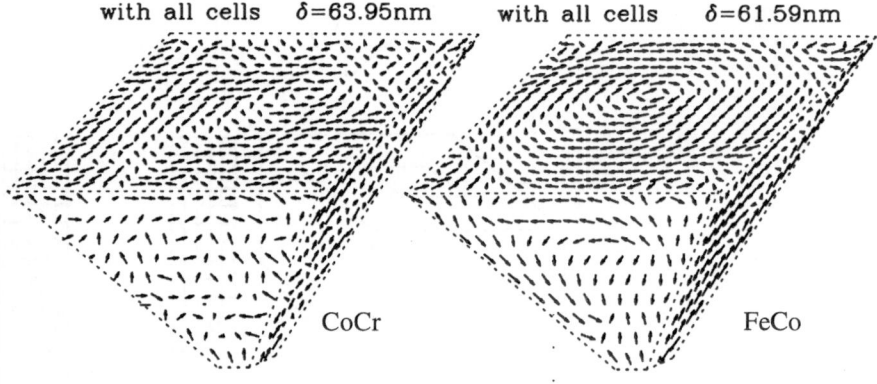

Fig. 4.18 Three-dimensional domain structure of pyramid MFM tip with all cells in model A included **a** CoCr; **b** FeCo

In Model A, pyramid-shaped MFM tips made by CoCr or FeCo are calculated and analyzed. Model A has a total number of $20 \times 10 \times 20$ micromagnetic cells, where the tip is located in the region within $19 \times N_y \times 19$ cells. Each cell has a dimension $10 \times 14.14 \times 10$ nm, labeled as $a \times c \times a$ in Fig. 4.17a; thus the side surfaces of the pyramid are the (111) crystal surfaces of silicon before the deposition of magnetic material. The simulation parameters are listed in Table 4.4.

Since the tip is usually initially magnetized vertically, the initial moments are set in the z-direction. Figure 4.18a shows the 3-D domain structure of the static state in a CoCr pyramid-shaped tip with all three types of cells included in the micromagnetic model, where the dipole position $\delta = 63.95$ nm in CoCr tip with all cells included is lower than the dipole position $\delta = 76.10$ nm in a simpler model with only cuboid cells included. The vortex magnetization structure is more obvious in the pyramid tip with all cells included, which is attributed to the contributions of the large demagnetizing field and shape anisotropy field of triangular prism cells and quadrangle pyramid cells. In Fig. 4.18b, it can be seen that a pyramid tip made of the soft magnetic material FeCo with large saturation magnetization M_s has much stronger vortex magnetization structure than the tip coated with a hard magnetic material CoCr, which is attributed to large demagnetizing field.

The difference between the tip field distribution calculated by micromagnetic model and point probe model can be found at different tip–sample distances $d_{t.s.}$, as seen in Fig. 4.19. In the CoCr tip, the tip field calculated by the micromagnetic model has a narrower distribution (PW50) compared to the dipole field, due the complicated

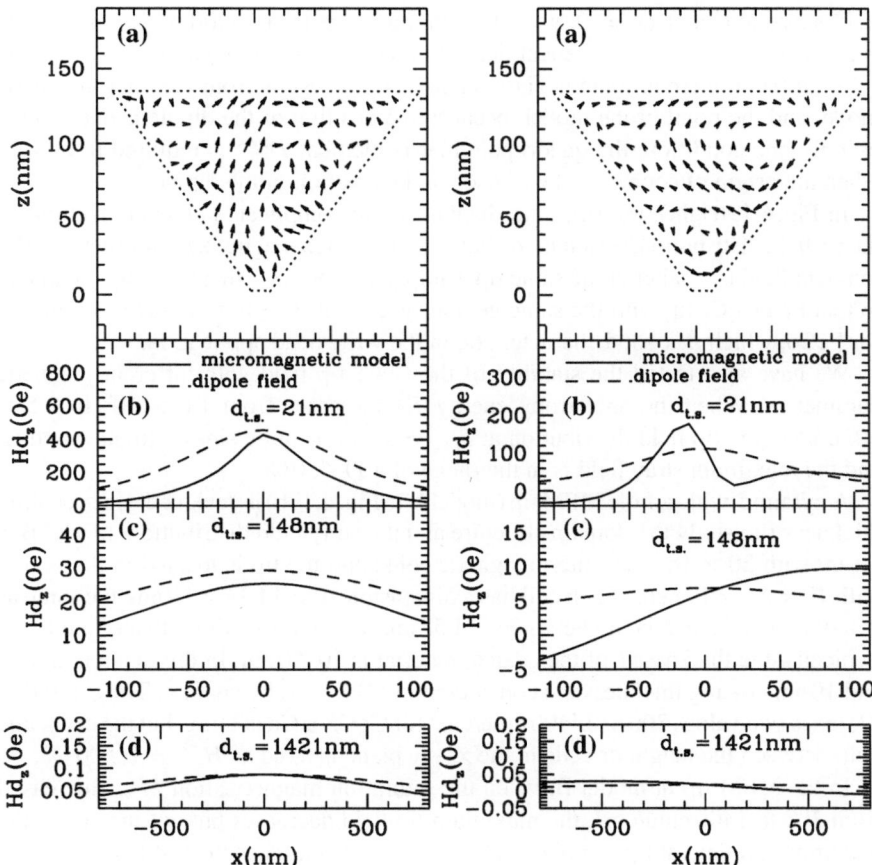

Fig. 4.19 Moments and field distribution of static state in CoCr (*left*) and FeCo (*right*) pyramid tip in model A. **a** Moments distribution in $x - z$ plane; (**b–d**) comparison of the MFM tip field along the x-axis calculated by the micromagnetic model or by the point probe model, with tip–sample distance $d_{t.s.} = 21, 148, 1, 421$ nm, respectively; © [2010] IEEE. Reprinted, with permission, from Ref. [13]

domain structure in the tip. Figure 4.19(left) also shows that the calculated field of this CoCr tip exhibits symmetry within the $x - y$ plane, which is very similar to that of the dipole field. The field distributions have little difference in x and y directions, which is not shown here.

The accuracy of the theory of effective dipole's position in Eq. (4.5) can be tested. In the CoCr tip, with increasing $d_{t.s.}$, the difference of the stray field at $x = 0$ between the two methods reduces from 20% ($d_{t.s.} = 21$ nm) to 10% ($d_{t.s.} = 148$ nm). When tip–sample distance $d_{t.s.}$ is more than ten times the tip height, e.g. $d_{t.s.} = 1, 421$ nm, the difference between the two models can be ignored (less than 2%); furthermore, the tip field distribution calculated by the micromagnetic model equals the dipole field near the center ($x = 0$, $y = 0$) within a range from -200 to $200\,$nm, which is a verification of our theoretical derivation and the micromagnetic model. When $|x| > 200\,$nm, there is some deviation between the results from the micromagnetic model and the point probe model, because the position of the dipole is determined by only one condition: the quadrupole matrix element D_{33}^{new} is required to be zero when the origin of coordinates is chosen at the position of the dipole.

In Fig. 4.19(right), the FeCo tip displays more complicated characterizations in the tip field distribution, which is attributed to the vortex magnetization structure. The remnant field is smaller at the same tip–sample distance $d_{\text{t.s.}}$ in a FeCo tip compared to that in a CoCr tip with the same geometry, even if the saturation is much higher for FeCo, which is a good characteristic in the write head application.

We have also tested the stability of the CoCr tip field with different magnetic parameters. When the anisotropy energy K_1 increases from 1.0×10^6 to $2.5 \times 10^6\,$erg/cc, the tip field distribution at $d_{\text{t.s.}} = 21\,$nm is quite similar to one another, and the maximum stray field is in the range of 400–550 Oe.

A pyramid-with-a-hole MFM tip (model B in Fig. 4.17b) is designed with a coating thickness t to study the domain structure and the stray field distribution. Model B is set up with $20 \times 18 \times 20$ micromagnetic cells; and the tip is located in the region with $19 \times N_y' \times 19$ cells. Each cell has a dimension $10 \times 14.14 \times 10\,$nm, the same as that of model A; and the tip height is $254.52\,$nm, which is two times that of model A. Typically, the thickness t of the coated material is 10–50 nm. In this section, 50, 30 and 10 nm coating thickness are considered in MFM tips by coating. The saturation M_s takes two values 300 or $1,400\,$emu/cm^3 for CoCr or Co coating, but the exchange field between the neighbor cells in the $x - y$ plane is fixed as $H_e^{xy} = 1,500\,$Oe.

In the MFM tip in model B, when the saturation magnetization M_s is increased from 300 to 1400 emu/cm^3, the maximum tip field decreases but not increases: the maximum tip field at tip–sample distance $d_{\text{t.s.}} = 21\,$nm is decreased from approximately 700 Oe to less than 200 Oe in a tip with $t = 50\,$nm coating, from about 550 to 400 Oe in a tip with $t = 30\,$nm coating, and from nearly 350 to 250 Oe in a tip with $t = 10\,$nm coating. This phenomenon corresponds to the appearance of the vortex magnetic structures with a higher saturation $M_s = 1,400\,$emu/cm^3, where the magnetic moments at the tip end are aligned horizontally to form a vortex, as seen in Fig. 4.20b; correspondingly, the tip field distribution is asymmetric and the field magnitude is small, as plotted in Fig. 4.20d, respectively.

In model B of the pyramid-with-a-hole tip, when the tip–sample distance equals 1,421 nm (about 5.6 times the tip size), big difference (20–50%) still can be found between the tip field calculated by the micromagnetic model and the dipole field, which is the case when the dipole position is located in the "hollow" nonmagnetic area in the middle of the tip, as seen in Fig. 4.20e–f. This behavior is different from the situation as in model A of the pyramid-shaped tip.

In previous models of MFM tips, the orientation distribution of anisotropy field is isotropic thus the $\mathbf{H}_k$ is random in space. Controlling the magnetic properties of the coating magnetic thin film on the MFM tip has been reported to be effective in reducing the magnetic volume and improving the resolution [14]. Further studies on the effect of the magnetic anisotropy, such as perpendicular magnetic anisotropy (PMA),

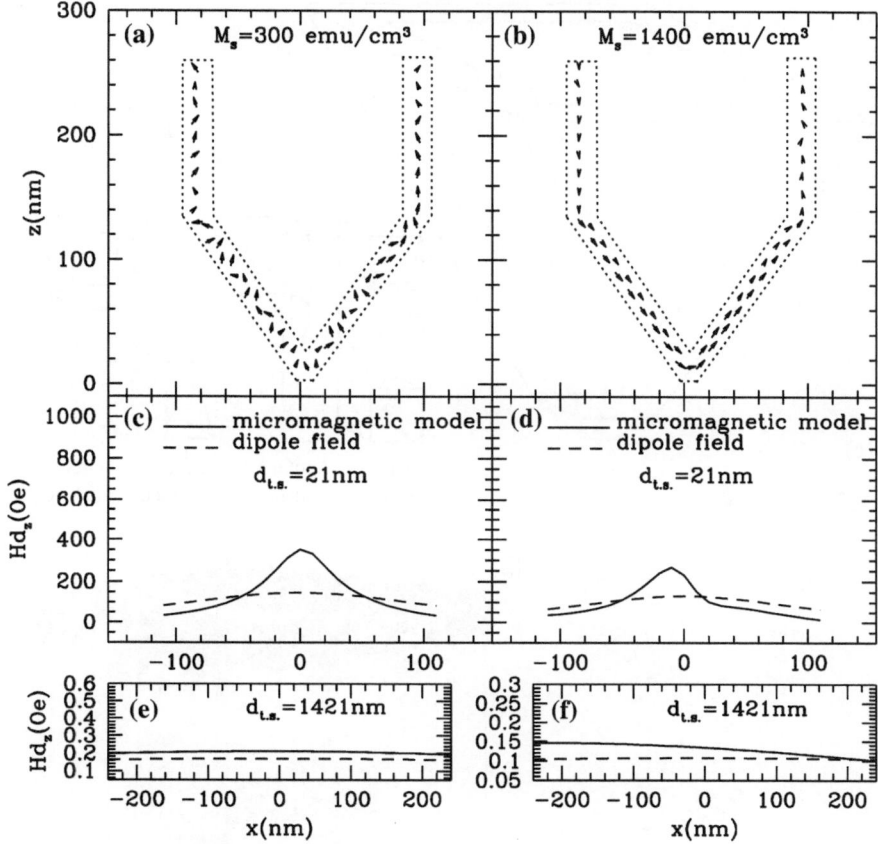

Fig. 4.20 Magnetic moments and dipole field distribution at tip–sample distance $d_{t.s.} = 21$ and 1,421 nm for MFM tips in model B with a 10 nm-thick CoCr or Co coating of different saturations; © [2010] IEEE. Reprinted, with permission, from Ref. [13]

random magnetic anisotropy (RMA) and in-plane magnetic anisotropy (IMA), of the tip-coating on the image resolution and signal are studied.

An accurate 3-D micromagnetic model of a pyramid MFM tip, with a cone angle of 37°, tip height of 120 nm, and a CoCrPt thin film coating thickness of 7.6 nm, was set up to calculate the domain structure and the stray field of the tip [15]. A CoCrPt disk medium, with a bit size of 24×32 nm^2 and film thickness of 8 nm [16], was chosen for image simulation and the studies of the tip–sample interaction. Both the tip and medium model use a regular mesh with a cell size of $8 \times 8 \times 8$ nm^3.

In the tip, both the inner and outer surfaces are smooth, thus many polyhedron cells have to be used for inclined surface [4]. Figure 4.21a presents the shapes of the polyhedron cells, whose demagnetizing matrices can be calculated analytically by summing over the contributions of surfaces. The CoCrPt thin film is polycrystalline, therefore there are magnetic grains distributed on the surface of the tip, as

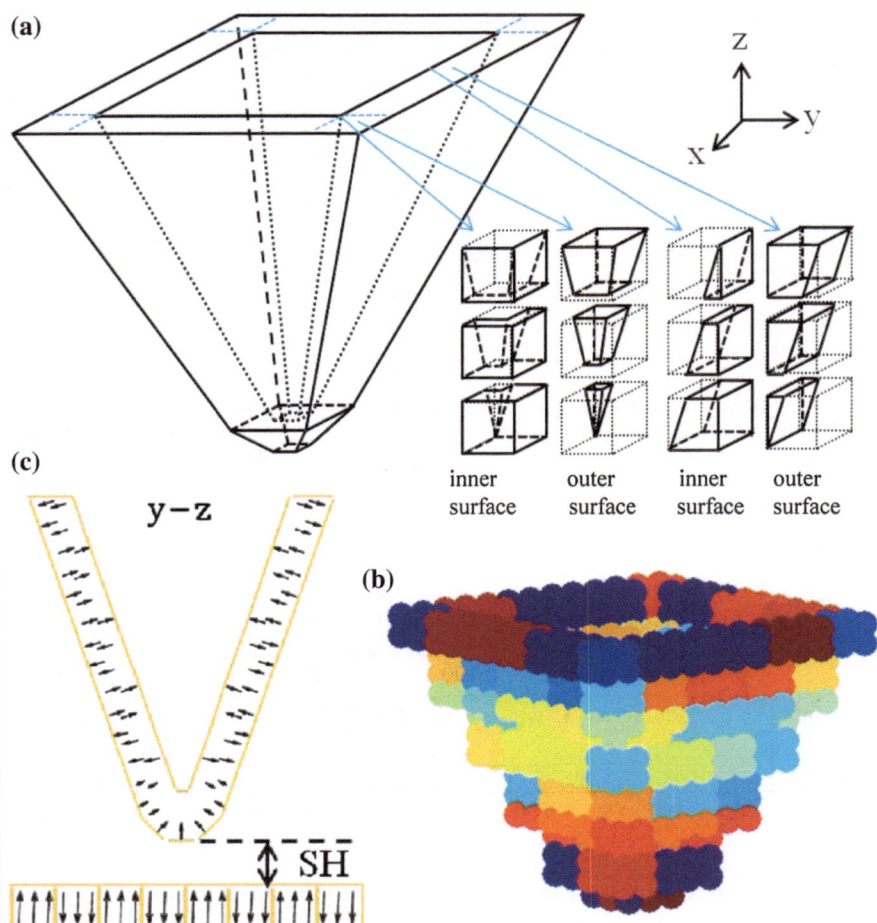

Fig. 4.21 A 7.6 nm CoCrPt-coating MFM tip with perpendicular magnetic anisotropy (PMA) **a** micromagnetic model with smooth inner and outer surface, with eight different types of polyhedron cells outer than the cubic cell; **b** color map of magnetic grains, where exchange between neighbor cells are large in the same grain and small in different grains; **c** magnetization distribution in the cross section of a PMA tip, with the moments in medium included in the model

shown in Fig. 4.21b, with a volume-average grain size of 8.8 nm, where the intra-grain and inter-grain exchange can be treated separately. Thus in the static domain structure shown in Fig. 4.21c, the magnetic moments belong to the same grain are stick together, but the moments in different grains can be antiparallel.

In the tip, the uniaxial crystalline anisotropy field constant is $H_k^{tip} = 1.8$ T. The saturation magnetization M_s is 700 emu/cc, the intra-grain exchange constant is $A_1^* = 3.0 \times 10^{-6}$ erg/cm and the inter-grain exchange constant should be in the range of $A_2^* = 0.3 - 1.0 \times 10^{-6}$ erg/cm for a stable image. For the MFM tip with PMA

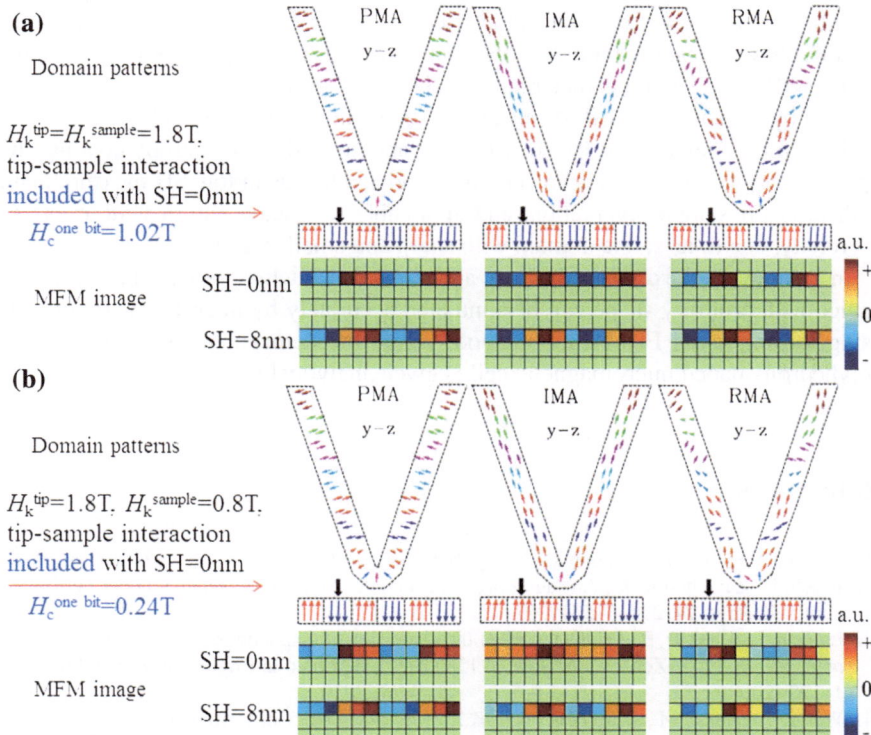

Fig. 4.22 Static domain states in tips ($A_2^* = 0.3 \times 10^{-6}$ erg/cm) and disk medium ($A^* = 0.3 \times 10^{-6}$ erg/cm); and respective simulated twelve MFM image points (along one scan line) calculated with the tip–sample interaction included in two media of **a** $H_k^{\text{sample}} = 1.8$ T and **b** $H_k^{\text{sample}} = 0.8$ T. (In domain patterns, the thick *black arrows* point the position of tip center, and different colors of the arrows in the tip stand for different grains and those in the medium stand for different bits)

coating, the tip field corresponding to the domain structure in Fig. 4.21c is very sharp and small, due to demagnetizing field cancelation of neighbor cells. At a tip–sample distance $d_{\text{t.s.}} = 2$ nm, the peak perpendicular field $\max\{H_y\}$ is 199 Oe, and the pulse width of H_y near the threshold (0–10 Oe) is only 16 nm for the PMA tip.

The domain patterns in MFM tip and disk media are simulated without or with tip–sample interaction. In a hard magnetic media with $H_k^{\text{sample}} = 1.8$ T (the corresponding coercivity of one bit is 1.02 T), it is found that the PMA, IMA or RMA tips have little influence on magnetic moments in the disk medium; also the disk medium has little influence on the moment distribution of PMA tip and IMA tip, but it seriously disturbs the magnetization pattern in the RMA tip. In a softer medium, where $H_k^{\text{sample}} = 0.8$ T (the corresponding coercivity of one bit is 0.24 T), at very low scan height (SH) SH=0 nm, the moments in the medium right under the tip are totally reversed by the strong stray field of IMA tip; The stray field of the disk medium has little influence on the moments distribution of the PMA tip and IMA tip, but it has big influence on the moments distribution of the RMA tip at both scan heights SH=0 and 8 nm. These characteristics of the PMA, IMA or RMA tips can also be verified by the simulated MFM images, as shown in Fig. 4.22 respectively.

Therefore, compared to the tip with in-plane or random anisotropy coating, the tip with perpendicular anisotropy coating is advantageous in low SH measurement, with little tip–sample interaction, and applicable for media with a wide range of coercivity, due to its stable domain with sharp and small stray field.

In a summary, the domain structure and dynamic switching process in a magnetic device with arbitrary shape can be simulated accurately by micromagnetic models using the improved FDM–FFT method, where the demagnetizing matrix of any polyhedron-shaped micromagnetic cell is given analytically.

References

1. Brown, W.F., Jr.: Micromagnetics. Wiley, New York (1963)
2. Bloch, F.: Zur Theorie des Austauschproblems und der Remanenzerscheinung der Ferromagnetika. Z. Phys. **74**, 295–335 (1932)
3. Landau, L., Lifshitz, E.: On the theory of the dispersion of magnetic permeability in ferromagnetic bodies. Phys. Zeitsch. der Sow. **8**, 153 (1935), reprinted in English by Ukr. J. Phys. **53**, 14–22 (2008)
4. Wei, D., Wang, S.M., Ding, Z.J., Gao, K.Z.: Micromagnetics of ferromagnetic nano-devices using fast Fourier transform method. IEEE Trans. Magn. **45**(8), 3035–3045 (2009)
5. Wang, S.M., Wei, D., Gao, K.Z.: Limits of discretization in computational micromagnetics. IEEE Trans. Magn. **47**(10), 3813–3816 (2011)
6. Rave, W., Ramstock, K., Hubert, A.: Corners and nucleation in micromagnetics. J. Magn. Magn. Mater. **183**, 329 (1998)
7. NIST: μMAG—Micromagnetic Modeling Activity Group. NIST Center for Theoretical and Computational Materials Science. http://www.ctcms.nist.gov/ rdm/mumag.org.html (2000). Accessed 20 June 2011
8. Iwasaki, S., Nakamura, Y., Ouchi, K.: Perpendicular magnetic recording with a composite anisotropy film. IEEE. Trans. Magn. **15**(6), 1456–1458 (1979)
9. Wang, S.M., Wei, D., Gao, K.Z.: Reversal properties of write head at extremely high density. IEEE. Trans. Magn. **45**(10), 3672–3675 (2009)
10. Wang, S.M., Wei, D., Gao, K.Z.: Initial permeability and dynamic response of FeCo write pole. IEEE Trans. Magn. **46**(6), 1951–1954 (2010)
11. Martin, Y., Wickramasinghe, H.K.: Magnetic imaging by "force microsopy" with 1000 Å resolution. Appl. Phys. Lett. **50**, 1455–1457 (1987)
12. Hartmann, U.: The point dipole approximation in magnetic force microscopy. Phys. Lett. A **137**, 475–478 (1989)
13. Li, H.J., Wang, Y., Wang, S.M., Zhong, H., Wei, D.: Micromagnetic analysis of effective magnetic dipole position in magnetic force microscope tip. IEEE. Trans. Magn. **46**(7), 2570–2578 (2010)
14. Grütter, P., Rugar, D., Mamin, H.J., Castillo, G., Lambert, S.E., Lin, C.-J., Valietta, R.M., Wolter, O., Bayer, T., Greschner, J.: Batch fabricated sensors for magnetic force microscopy. Appl. Phys. Lett. **57**, 1820 (1990)

15. Li, H.J., Wei, D., Piramanayagam, S.N.: Micromagnetic studies on resolution limits of MFM tips with different magnetic anisotropy. J. Appl. Phys. **111**, 07E309 (2012)
16. Piramanayagam, S.N., Ranjbar, M., Tan, E.L., Tan, H.K., Sbiaa, R., Chong, T.C.: Enhanced resolution in magnetic force microsopy using tips with perpendicular magnetic anisotropy. J. Appl. Phys. **109**, 07E326 (2011)

Chapter 5
Erratum to: Microstructure and Hysteresis Loop

Dan Wei

Erratum to: Microstructure and Hysteresis Loop, DOI:10.1007/978-3-642-28577-6_3

The original version of Chapter 3 unfortunately contained a mistake. The presentation of Fig. 3.6b was incorrect. The corrected figure is given below.

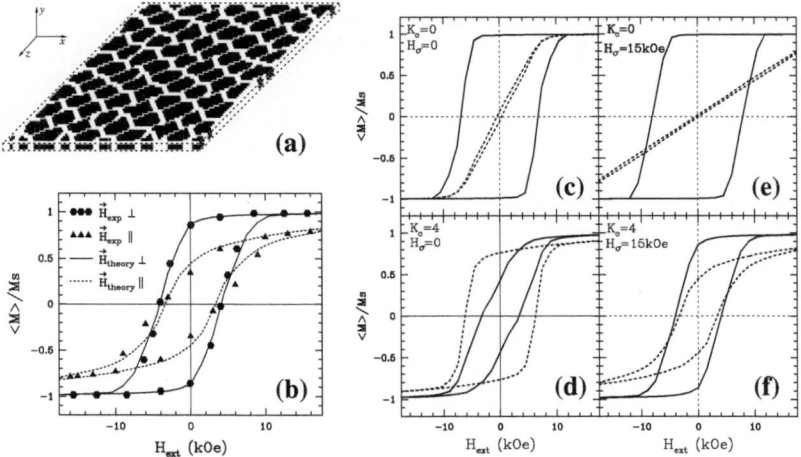

Fig. 3.6 Simulation of FePt media. **a** 3-D view of microstructure; **b** simulated perpendicular and in-plane loops compared with experiment; **c–f** effects of tetragonal anisotropy term $K_c = 0, 4K_{u1}$ and magnetostriction field $H_\sigma = 0, 15$ kOe on the M–H loops

The online version of the original chapter can be found at DOI:10.1007/978-3-642-28577-6_3.

D. Wei (✉)
Key Laboratory of Advanced Materials, Department of Materials Science and Engineering,
Ministry of Education, Tsinghua University, Beijing, 100084, China
e-mail: weidan@tsinghua.edu.cn

D. Wei, *Micromagnetics and Recording Materials*, SpringerBriefs in
Applied Sciences and Technology, DOI: 10.1007/978-3-642-28577-6_5,
© The Author(s) 2012

Index

D. Wei, *Micromagnetics and Recording Materials*, SpringerBriefs in
Applied Sciences and Technology, DOI: 10.1007/978-3-642-28577-6,
© The Author(s) 2012